the
UNIVERSITY
of
GREENWICH

CROP–SOIL SIMULATION MODELS
APPLICATIONS IN DEVELOPING COUNTRIES

Crop–Soil Simulation Models
Applications in Developing Countries

Edited by

Robin Matthews
and
William Stephens
Institute of Water and Environment
Cranfield University
Silsoe
UK

CABI *Publishing*

CABI *Publishing* is a division of CAB *International*

CABI Publishing
CAB International
Wallingford
Oxon OX10 8DE
UK

CABI Publishing
10 E 40th Street
Suite 3203
New York, NY 10016
USA

Tel: +44 (0)1491 832111
Fax: +44 (0)1491 833508
Email: cabi@cabi.org
Web site: www.cabi-publishing.org

Tel: +1 212 481 7018
Fax: +1 212 686 7993
Email: cabi-nao@cabi.org

A catalogue record for this book is available from the British Library, London, UK.

Library of Congress Cataloging-in-Publication Data
Crop–soil simulation models: applications in developing countries/edited by Robin Matthews and William Stephens.
p. cm.
Includes bibliographical references.
ISBN 0-85199-563-2 (alk. paper)
1. Crops and soils—Computer simulation. 2. Crops—Computer simulation. 3. Soils—Computer simulation.
4. Crops and soils—Mathematical models. 5. Crops—Mathematical models. 6. Soils—Mathematical models.
I. Matthews, Robin B. II. Stephens, William, Ph. D.

S596.7.C72 2002
630'.1'13--dc21   2001052815

ISBN 0 85199 563 2

Typeset by Wyvern 21 Ltd, Bristol.
Printed and bound in the UK by Cromwell Press, Trowbridge.

Contents

Contributors

A. Graves, Institute of Water and Environment, Cranfield University, Silsoe, Bedfordshire MK45 4DT, UK.

T. Hess, Institute of Water and Environment, Cranfield University, Silsoe, Bedfordshire MK45 4DT, UK.

R. Matthews, Institute of Water and Environment, Cranfield University, Silsoe, Bedfordshire MK45 4DT, UK.

T. Middleton, Institute of Water and Environment, Cranfield University, Silsoe, Bedfordshire MK45 4DT, UK.

W. Stephens, Institute of Water and Environment, Cranfield University, Silsoe, Bedfordshire MK45 4DT, UK.

R. Wassmann, Fraunhofer Institute for Atmospheric Environmental Research, Kreuzeckbahnstrasse 19, 82467 Garmisch-Partenkirchen, Germany.

Abbreviations

AEGIS	Agricultural and Environmental Geographic Information System
AMMI	Additive Main effects and Multiplicative Interaction
ANOVA	Analysis of variance
APSIM	Agricultural Production Systems Simulator
AWS	Automatic weather station
BBF	Broadbed-and-furrow
CAL	Computer-assisted learning
CAP	Common Agricultural Policy
CATIE	Centro Agronomic Tropical de Investigacion y Ensenanza
CERES	Crop Environment Resource Synthesis
CIP	International Potato Centre
CLUES	Centre for Land Use Studies
CTI	Computers in Teaching Initiative
DAD	(South African) Department of Agricultural Development
DFID	Department for International Development
DNDC	Denitrification and decomposition
DSS	Decision support system
DSSAT	Decision Support System for Agrotechnology Transfer
ENSO	The El Niño-Southern Oscillation
EPIC	Erosion Productivity Impact Calculator
EPIPRE	Epidemic prevention
ES	Expert system
EU	European Union
FAO	Food and Agriculture Association

FARMSCAPE	Farmers, Advisers and Researchers Monitoring Simulation, Communication and Performance Evaluation
FRS	Fertilizer recommendation system
FSR	Farming systems research
GCMs	General circulation models
GFDL	General Fluid Dynamics Laboratory
GHGs	Greenhouse gases
GIS	Geographical information system
GISS	Goddard Institute of Space Studies
G×E	Genotype by environment
HRI	Horticultural Research International
IBSNAT	International Benchmark Sites Network for Agrotechnology Transfer
ICASA	International Consortium for Agricultural Systems Applications
ICRISAT	International Crops Research Institute for the Semi-Arid Tropics
IMS	Irrigation management services
IPM	Integrated pest management
IRRI	International Rice Research Institute
IWR	Irrigation Water Requirements
LDCs	Less-developed countries
LINTUL	Light Interception and Utilization
LP	Linear programming
LUT	Land-use type
MAFF	Ministry of Agriculture Fisheries and Food
MARS	Monitoring Agriculture with Remote Sensing
MDS	Minimum data set
MERES	Methane Emission from Rice Ecosystems
METs	Multi-environment trials
NARCs	National Agricultural Research Centres
NGO	Non-governmental organization
NR	Natural resources
OR	Operations research
QTL	Quantitative trait loci
R&D	Research and development
RLWR	Root length/weight ratio
REPOSA	Research Programme for Sustainability in Agriculture
RIL	Recombinant inbred line
RUE	Radiation use efficiency
SARP	Simulation and Systems Analysis for Rice Production
SASEX	(South African) Sugar Association Experiment Station
SL	Sustainable Livelihoods
SLA	Specific leaf area
SOI	Southern Oscillation Index

SOLUS	Sustainable Options for Land Use
SWB	Soil water balance
UKMO	United Kingdom Meteorological Office
UNED	Universidad Estatal a Distancia
USAID	United States Agency for International Development
WARDA	West African Rice Development Association
WUE	Water use efficiency

Preface

From 1990 until 1999, the United Kingdom's Department for International Development (DFID) funded work on developing a suite of models to address problems relating to crop production in the semiarid tropics, specifically the evaluation of rain-water harvesting, and maintenance of soil fertility. This work resulted in the PARCH, PARCHED-THIRST, SWEAT and EMERGE models. However, a number of subsequent studies (e.g. Fry, 1996; Stephens and Hess, 1996; Kebreab et al., 1998) showed that uptake and use of these models was limited to non-existent. This has led to questions being asked as to whether crop simulation modelling and systems analysis approaches have any contribution to make in addressing problems in developing countries.

A general weakness of all the models was that a clear definition of who potential users were had received scant attention. The models were developed to support the solution of problems in natural resources management, but this was not in response to a known and well-articulated demand from potential users in the natural resources sector of developing countries. The study by Stephens and Hess (1996) identified the first limitation to the uptake of a model as the inability of a potential user to be able to perceive a relevance of the model to his/her work, or lack of appreciation of what the model could be used for. All three studies emphasized the need for continued support to end-users of the models if there is to be uptake. Both of these points have been recognized by the Natural Resources Systems Programme (NRSP) of DFID, and in October 1999, a workshop funded by NRSP was held at IACR-Rothamsted to review the current status of the PARCH suite of models and crop models in general, and to discuss options for taking DFID-funded modelling activities forward, with particular emphasis on the application of systems approaches to contribute to the solution of real-world problems in developing countries. This book is a progression from discussion points raised in the wrap-up session of this workshop. The purpose of the work was to make a thorough review of the literature to identify past and current applications of crop–soil simulation models in general, identify the limitations of such models, characterize groups of end-users of the models, and to attempt a synthesis of where such models might be useful in the future in contributing to system-based, poverty-oriented research projects in developing countries.

Many people have contributed to the ideas in this book in many discussions over several years. Prominent among these are Professor Tony Hunt, at the University of Guelph, Canada; Professor Jim Jones and Professor Gerrit Hoogenboom, at the University of Gainesville, Florida; Dr Walter Bowen, IFDC; Professor Martin Kropff, Wageningen Agricultural University, The Netherlands; Dr P.K. Aggarwal, IARI, India; Dr Ino Lansigan, UPLB, The Philippines; Dr Attachai Jintrawet, Thailand; and Dr Kevin Waldie, University of Reading, UK. We are also grateful to the participants in the Rothamsted workshop in October 1999 for the useful discussions that set us on the route to writing this book – Dr John Gaunt, IACR-Rothamsted (the workshop organizer); Dr Georg Cadisch, Wye College; Dr Neil Crout, University of Nottingham; Dr John Gowing, University of Newcastle; Mr Gerry Lawson, Institute of Terrestrial Ecology; Mr Frans-Bauke van der Meer, Silsoe Research Institute; Dr Robert Muetzelfeldt, University of Edinburgh; Dr Lester Simmonds, University of Reading; Dr Terry Thomas, University of Wales, Bangor; Dr Geoff Warren, University of Reading; Dr Ermias Kebreab Weldeghiorghis, University of Reading; and Dr Damion Young, University of Newcastle.

Substantial use was also made of Internet discussion groups to obtain information and views from the international community involved in modelling agricultural systems. Principally, these were the AGMODELS listserver (AGMODELS-L@UNL.EDU), the DSSAT listserver (DSSAT@LIST-SERV.UGA.EDU), the ESA-AGMODELS listserver (ESA-AGMODELS@ESA.UDL.ES), and the FAO-AGROMET listserver (AGROMET-L@MAILSERV.FAO.ORG). We are grateful to all subscribers of these groups who responded to our questions, and have attempted to give credit to the source where we have included these comments in this book.

This publication is in large part an output from a programme development assignment funded by DFID for the benefit of developing countries. The views expressed are not necessarily those of DFID.

A condensed version of Part 1 has been published in *Advances in Agronomy* (Matthews *et al.*, 2002).

Introduction

1

Robin Matthews

Institute of Water and Environment, Cranfield University, Silsoe, Bedfordshire MK45 4DT, UK

Arable agriculture is a major way in which people interact with the natural resource base in developing countries; this may not always be to the long-term benefit of either, particularly if cropping practices are suboptimal or inappropriate. Traditional agronomic research has made remarkable advances in recent years in improving some of these practices, but new tools being developed, such as crop and soil simulation models with their ability to integrate the results of research from many different disciplines and locations, offer a way of improving the efficiency and/or reducing the cost of some of this research. Since research organizations cannot afford to generate technologies that are inappropriate, more use is being made of systems methods to ensure that research is relevant (Goldsworthy and Penning de Vries, 1994). Their use in a research programme has the potential to increase efficiency by emphasizing process-based research, rather than the study of site-specific net effects. This is particularly attractive in developing countries, where scarce resources may limit effective agricultural research.

There is also an increasing need to understand how agricultural systems interact with other segments of society. The population of the world is increasing by over 70 million per year, and it is likely that there will be 1.8 billion more people in the world by 2020 (Pinstrup-Andersen *et al.*, 1999). To meet the demand for food from this increased population, the world's farmers need to produce 40% more grain by 2020. Moreover, if certain climate change scenarios come to pass, agricultural production in some areas may decrease. There are many cases of land degradation, and a lack of new land that can be brought into agricultural production. How can productivity be increased while ensuring the sustainability of agriculture and the environment for future generations? Decision makers need information supplied by research to make informed choices about new agricultural technologies and to devise and implement policies to enhance food production and sustainability. Ultimately, however, it is the farmer who makes the final choices about acceptance of a new technology or method. Policy makers need to understand the impacts of their decisions on the wellbeing of farm households, on the natural resource base, and on the regional or national economy. The users of information generated through research and encapsulated in models are not just farmers but decision makers at all levels in the public and private sectors.

There are many types of models that have been published describing various aspects of agricultural production systems, and it is all too easy to be overwhelmed by their sheer numbers. However, as this book evolved from discussions on the crop modelling work already funded by DFID and how this might be taken forward, we have, therefore, restricted our focus to crop simulation models, which may or may not have components describing soil processes and pests and diseases (Penning de Vries, 1990). We have adopted the definition of Sinclair and Seligman (1996) that a crop model is 'the dynamic simulation of crop growth by numerical integration of constituent processes with the aid of computers'. More specifically, this implies a computer program describing the dynamics of the growth of a crop (e.g. rice, wheat, maize, groundnut, tea, etc.) in relation to the environment, operating on a time-step an order of magnitude below the length of a growing season, and with the capacity to output variables describing the state of the crop at different points in time (e.g. biomass per unit area, stage of development, yield, canopy N content, etc.). We have not generally included models that only predict some final state such as biomass or yield (Whisler et al., 1986). Nevertheless, we have permitted ourselves to deviate occasionally and consider models and their applications that might be outside this definition, but only if we feel that there are lessons to be learned that are relevant to the way ahead for crop and/or soil simulation modelling in the context of natural resources systems research.

In a task of this kind, it becomes necessary to classify model applications in order to provide the basis for some kind of meaningful discussion. There are many different ways that the uses of models can be classified: Passioura (1996), for example, classifies models into two groups – scientific models (i.e. helping with understanding) and engineering (i.e. applying science to solve a problem). Mindful of the broad groups of end-users of crop simulation models, we have expanded this classification, and have divided model applications into: (i) those used as tools by researchers, (ii) those used as tools by decision-makers, and (iii) those used as tools by those involved in education, training and technology transfer. We are the first to recognize that this is not a perfect classification and that there is bound to be overlap between the groups, but have found it to be a useful way of thinking about common characteristics of models from the point of view of the people who will be using them. Where a particular application falls into more than one classification, we generally discuss it under both headings, with the focus on the aspects relevant to each classification.

We have not attempted to review every single instance of a crop model application, as that would require considerably more time than we had available. Instead, we have attempted to cover all the broad types of uses to which crop models have been put, and have used as many examples of each as possible to illustrate the use of models in that area. We recognize that there will probably be many good examples of model applications that we have not included; we apologize to the people involved and hope that they can appreciate that it is only space and time that prevented us from doing so. We have also focused on applications of crop models in developing countries, although we have drawn substantially on experiences with tropical crops in Australia, as many of the examples there are relevant to possible applications of systems analysis techniques in other tropical countries. We have not generally included examples from temperate agricultural systems, except where we feel there were interesting lessons learned that had some relevance to agriculture in developing countries.

We also recognize that there is a certain element of unavoidable bias in such a review towards instances where models have been successfully applied. Cases where models have failed or have been unsuccessfully applied are generally not reported in the literature. However, we make no apologies for this bias, in the same way as a plant breeder is not required to apologize for the 99% or more of his material that are 'failures'. Just as progress in plant breeding is made with the proportion of individuals that are 'successful', the aim of this study is to as impartially as possible identify areas where models have been applied successfully, so that future modelling activities can be focused in those areas. Indeed, we would argue that the notion of a model being a 'success' or a 'failure' is somewhat meaningless, anyway – in research, the most useful model is often the one that fails as it can point the way to new thinking and research (Seligman, 1990). On the other hand, we do recognize that there is a cost to research that has not produced results, and with this in mind we have attempted to make appraisals of the limitations of the models used where possible, and have discussed constraints to their uptake and impact.

Before describing the various applications of crop simulation models that have been published, it is perhaps useful to summarize the history of crop modelling to provide some perspective. Sinclair and Seligman (1996) have given an excellent overview of developments, drawing parallels between the growth and development of crop simulation models and human beings.

They describe how the *infancy* stage began after the birth of crop modelling more than 35 years ago with the advent of the mainframe computer in the 1960s (Bouman *et al.*, 1996). The first steps for crop models were models designed to estimate light interception and photosynthesis in crop canopies (e.g. Loomis and Williams, 1962; de Wit, 1965). These were

relatively simple models, but they provided for the first time a way of quantitatively and mechanistically estimating attainable growth rates of various
crops. They showed that the potential yield of a crop could be defined in
terms of the amount of solar radiation energy available for the accumulation of chemical energy and biomass by plants. The *juvenile* stage that
followed in the 1970s seemed to open up wide areas of research, and led
to the development of so-called 'comprehensive' models mainly aiming at
increasing understanding of the interactions between the crop and the main
growth factors. This stage also coincided with rapid advances in equipment for field experimentation to provide data describing the various physiological processes that were incorporated into the models, which
inevitably led to an increase in their complexity. However, this complexity meant that the number of parameters required to describe the system
in detail increased dramatically. Errors in the values of these parameters
obtained from field experimentation often propagated through the model.
Other parameters could not be measured directly and had to be estimated.

The *adolescence* stage in the early 1980s saw a re-evaluation of the basic
concepts of crop modelling in the light of accumulated evidence. The first
of these was the assumption that the reductionist approach of increasing
the complexity of a model led to better models. It had become apparent
that much of the behaviour of a system could be captured by a few key
variables, with the inclusion of further variables only adding marginally to
the accuracy of the model, if at all. This led to the emergence of simplified
versions of the comprehensive models, or so-called 'summary models'
(Penning de Vries *et al.*, 1989), and, even more recently, still simpler 'parsimonious' models (e.g. ten Berge *et al.*, 1997b; Peiris and Thattil, 1998).
In these latter models, a system is modelled with only a few key variables
in an attempt to keep a model simple, both so that it is easily understood
by potential users, and also that its requirements for input data are markedly reduced. Nevertheless, a more detailed crop simulation model may often
be used to help develop the simpler model. The second re-evaluation was
of the original assumption that a universal model could be developed for
each crop, with the realization that the nature of the problem to be solved
dictated the nature of the most appropriate model to use. This brought a
move towards 'bespoke' models built with a specific purpose in mind.

The *maturity* phase in the 1990s brought a growing awareness of the
limitations of crop models and a better understanding of the nature of
these limitations, some of which we discuss in more detail in this book.
Many objections have been raised to the use of deterministic crop
growth models, ranging from lack of confidence in the method altogether
(e.g. Passioura, 1973; Monteith, 1981), through the problems of their data
requirements, the 'parameter crisis' (Burrough, 1989b), the stochastic nature
of the input data used (Burrough, 1989a), the fact that model results
necessarily pertain to single events which causes application problems in
spatially and temporally variable environments, to the complaint that the

models cannot reproduce the actual situation. On the positive side, however, there does seem to be general agreement across the board that the development of such models has brought benefits by providing the opportunity to formulate consistent quantitative statements on the behaviour of the systems under consideration, that the consequences of alternative options can therefore easily be made explicit, and as such, these models form a tangible basis for discussion.

In the following chapters, we would like to extend the human development analogy of Sinclair and Seligman (1996) and describe the first *employment* these crop models have had, and offer some thoughts about how their job prospects might develop from here, with particular focus on their relevance to agriculture in developing countries.

Part 1
Models as tools in research

Models as Research Tools

2

Robin Matthews

Institute of Water and Environment, Cranfield University, Silsoe, Bedfordshire MK45 4DT, UK

Crop simulation models were originally developed as research tools, and have probably had their greatest usefulness so far in being part of the research process. The advantages of integrating simulation modelling approaches into a research programme have often been stated – Seligman (1990), for example, lists the following uses of models in research:

- identification of gaps in our knowledge;
- generation and testing of hypotheses, and an aid to the design of experiments;
- determination of the most influential parameters of a system (sensitivity analysis);
- provision of a medium for better communication between researchers in different disciplines;
- bringing of researchers, experimenters and producers together to solve common problems.

Boote *et al.* (1996) see models as providing a structure to a research programme, and being particularly valuable for synthesizing research understanding and for integrating up from a reductionist research process, but point out that if the efficiency of research is to be increased, the model ling process must become a truly integrated part of the research activities. Experimentation and model development need to proceed jointly – new knowledge is used to refine and improve models, and models are used to identify gaps in our knowledge, thereby setting research priorities. Sinclair and Seligman (1996) make a similar point, seeing models as a way of setting our knowledge in an organized, logical dynamic framework, allowing identification of faulty assumptions and providing new insights.

They propose that models should be seen as aids to reasoning in research and teaching about the performance of a crop or the benefits of alternative management strategies.

An interesting example of the use of models to provide new insights into crop processes for future research to focus on is provided by Matthews and Stephens (1998b). During the development of a simulation model for tea (*Camellia sinensis*), it was found that temperature alone could not be used to simulate the large peak in tea production in September in Tanzania. Various potential mechanisms were evaluated, but the only one that was able to adequately explain this peak was the assumption that the growth of dormant shoots was triggered at the time of the winter solstice, allowing a large cohort of shoots to develop simultaneously and reach harvestable size at the same time (Fig. 2.1). The proposed mechanism, in which shoot dormancy was induced by declining photoperiod and released by increasing photoperiod, was also able to accurately simulate patterns of shoot growth in the northern hemisphere (Panda *et al.*, 2002). An experiment was planned to test the hypothesis, but was subsequently cancelled when the company that was to fund the work sold their interests in tea. The matter is not only academic – the large September production peak can often exceed factory processing capacity with a subsequent loss of harvested material. If the peak could be manipulated with, say, supplementary lighting to offset the photoperiod effect, a more even spread of production over the year might be obtained.

Many of the crop model applications discussed in the following pages involve an assessment of risk, so we feel that it is worthwhile to say a little

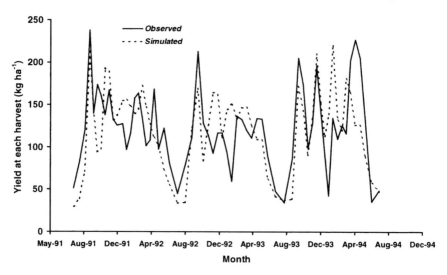

Fig. 2.1. Comparison of observed (——) yields of fully irrigated tea in Tanzania with yields simulated using the CUPPA-TEA model (----) (redrawn from Matthews and Stephens, 1998b).

at this stage about what risk is, what causes it, and how different people perceive risk in different ways. Risk and uncertainty are inherent in agriculture, particularly in tropical countries. The influence of weather, pests and diseases, and prices and costs, are all unknown in advance to varying degrees. This risk and uncertainty is important as it affects decision making at a number of levels – the same decision made in a low-risk environment may be totally inappropriate in a high-risk environment. Efficient agricultural management has much to do with the management of this risk, both at the household and regional levels. At the household level, a farming family may try to maximize income fluctuations over time; at the national level on the other hand, a government may try to ensure an adequate supply of food to the population in all sectors of society. Risk to crop production may be short term, such as fluctuations in climate or socio-economic conditions, or long term, such as degradation of soil fertility. Often production in the short term can be maximized, but sometimes it may be at the expense of an increase in resource degradation in the long term. Also, different people have different perceptions and time scales of risk – an individual farmer may be much more concerned about the risk of crop failure in the next season than the risk of long-term decline in soil fertility, whereas a government may be much more concerned about long-term harm to the environment. Wade (1991) classified farmers into three groups: (i) risk takers – those who aim for high productivity in a good year, and are prepared to accept crop failure or low yields in some years; (ii) risk avoiders – those who are prepared to sacrifice some yield in a good year as long as the risk of crop failure in poor years is minimized; and (iii) those who fall somewhere in between (i) and (ii). Commercial farmers with access to credit and who can afford high levels of inputs would generally fall into category (i), while subsistence farmers with little or no access to means to buffer year-to-year variability would generally fall into category (ii). With access to long sequences of historical weather data, crop models can be excellent tools for assessing the production variability associated with weather for various strategies (Thornton and Wilkens, 1998). Traditional field experiments to obtain the same assessment of risk associated with a particular strategy would be virtually impossible due to the time and cost involved.

In the following chapters describing applications of models as research tools, we have started at the level of the genotype, considering how models may contribute to the process of genotype improvement, then moved to the level of the whole crop, discussing applications aimed at understanding and improving crop management. We then progress to how individual crops fit into an overall farming system, looking at ways in which these systems can be optimized to meet certain goals and how they contribute to the livelihoods of the farmers involved. Finally, we consider how crop models have contributed to the policy-making process at the national and international levels.

Crop Genotype Improvement

3

Robin Matthews

Institute of Water and Environment, Cranfield University, Silsoe, Bedfordshire MK45 4DT, UK

The goal of any plant breeding programme is the development of new improved cultivars or breeding lines for particular target areas and for specific applications. In general, the time from initial selection of individual plants to the release of a new cultivar can take up to 10–15 years, and in most cases, improvements of only a few per cent are obtained for each new cultivar over current ones. Any new techniques of improving the efficiency of the improvement process are, therefore, of considerable interest to plant breeders. Chapman and Bareto (1996) have defined increasing the efficiency of plant breeding as increasing the rate of genetic gain, given particular levels of research resources and genetic variability.

The overall process of crop improvement can be subdivided into three phases – a planning and hybridization phase, a segregation and stabilization phase, and a line evaluation and release phase (Hunt, 1993). The personal time of a breeder can be allocated between these three phases in different ways, but a suggested allocation for a wheat improvement programme is shown in Table 3.1. Regardless of the type of crop improvement programme, most breeders consider the first phase, the design of a new genotype for a particular environment, the selection of parents with characteristics matching this design and the initial hybridization, to be of critical importance, with around 40% of their time being allocated to it (Hunt, 1993). Even with careful matching of parents, the chances of success depend on the numbers of lines evaluated each year. Thus, given the increasingly marginal returns from conventional breeding approaches, it is timely to seek more efficient methods that might help improve the efficiency of this phase.

The emergence of simulation models for a large number of crops provides tools that may be useful in helping to improve the efficiency of the

Table 3.1. Division of time between the different phases of activity in a self-pollinated crop breeding programme (Jensen, 1975).

Phase	Breeders' time (%)	Technicians' time (%)
1. Planning and hybridization	40	5
2. Segregation and stabilization	10	10
3. Line evaluation and release	50	85
	100	100

crop improvement process. Both Shorter *et al.* (1991) and Lawn (1994) stress the need for an integrated multidisciplinary approach between plant breeders, crop physiologists, and crop modellers. Cooper and Hammer (1996), summarizing the results of a workshop on plant adaptation and crop improvement held at the International Crops Research Institute for the Semi-Arid Tropics (ICRISAT), highlighted the use of models in crop improvement programmes in three main areas: (i) identification and evaluation of desirable plant characteristics to aid indirect selection methods; (ii) characterization of the target environments for different germplasm; and (iii) partitioning genotype × environment (G×E) interactions to increase the sensitivity of the analysis of variance of trial data. However, for crop simulation models to make significant contributions in these areas, there is a need for much collaborative research to be done between physiologists and plant breeders. Until this research is conducted, and the benefits of shifting resources into the systems approach can be weighed against reducing resources put into the conventional empirical approach, widespread acceptance of the new methods is unlikely (Hammer *et al.*, 1996a).

In the following sections in this chapter, work that addresses these three areas is summarized and discussed.

3.1 Identification and Evaluation of Desirable Plant Characteristics

Direct selection for crop yield is generally perceived as costly and inefficient because of its low heritability (White, 1998), despite being the main method of selection for superior germplasm up until the present time. Much effort, therefore, has gone into the identification of traits which breeders might select for to increase yield indirectly. Crop models offer a way in which various traits can be evaluated simply and easily. Varying only one plant parameter at a time while keeping the rest of the parameters constant is analogous to the creation of genetic isolines, something that requires a good deal of time and effort in reality. Although a single trait may be of interest, a combination of traits, or a crop ideotype, is more often sought.

The concept of designing a genotype with optimal characteristics for a particular set of conditions was first used by Donald (1968) who designed a small grain cereal for favourable environments. These ideas were subsequently expanded to develop a general ideotype applicable to cereals, grain legumes and oil seeds (Donald and Hamblin, 1983). The principal traits of this plant were an annual habit, erect growth, dwarf stature, strong stems, unbranched and non-tillered habit, reduced foliage, erect leaves, determinant habit, high harvest index and early flowering. Similarly, Cock *et al.* (1979) proposed a cassava ideotype with late branching, large leaves and long leaf life, based on a model that used weekly time intervals to simulate leaf development, crop growth and partitioning between roots and shoots. More recently, simulation studies helped in defining morphological characteristics of the 'New Plant Type' of rice currently being developed at the International Rice Research Institute (IRRI; Dua *et al.*, 1990; Dingkuhn *et al.*, 1991). Desirable traits were identified as: (i) enhanced leaf growth during crop establishment; (ii) reduced tillering; (iii) less foliar growth and enhanced assimilate export to stems during late vegetative and reproductive growth; (iv) sustained high foliar N concentration; (v) a steeper slope of N concentration from the upper to lower leaf canopy layers; (vi) expanded capacity of stems to store assimilates; and (vii) a prolonged grain filling period.

Because of their dynamic nature, crop simulation models also offer the opportunity to explore the effect of changing the rates of various physiological processes. Various cotton models, for example, have been used since 1973 to assess the effect on yield of traits including photosynthetic efficiency, leaf abscission rates and unusual bract types (e.g. Landivar *et al.*, 1983a, b; Whisler *et al.*, 1986). Landivar *et al.* (1983b) concluded that if photosynthetic efficiency was correlated with specific leaf weight, then most of the increased growth would go into the leaf with little overall effect on yield. Hoogenboom *et al.* (1988) used the BEANGRO model to investigate the effects of specific leaf area (SLA), root partitioning, rooting depth and root length/weight ratio (RLWR) on seed yield and water use efficiency (WUE) of common bean. Yields increased with an increase in rooting depth, root partitioning, increased RLWR, and increases in SLA up to 300 cm^2 g^{-1}, beyond which there was no increase. Boote and Jones (1988) performed a similar exercise with PNUTGRO, comparing the effects of 16 parameters on groundnut yield under rain-fed conditions over 21 years. Increasing canopy photosynthesis and the duration of the vegetative and reproductive phases both increased yields over 15%. Jordan *et al.* (1983) and Jones and Zur (1984) found that for soybean growing in a sandy soil, increased root growth was more advantageous than capacity for osmotic adjustment or increased stomatal resistance. By contrast, the GOSSYM model predicted that doubling stomatal resistance would lead to a 28% increase in yield, a conclusion subsequently supported by improved cultivars (Whisler *et al.*, 1986). Other examples of the use of crop simulation models to investigate the sensitivity of different genetic traits on yields are in soybean (Wilkerson *et al.*, 1983; Elwell *et al.*, 1987) and groundnut (Duncan *et al.*, 1978).

Determining the responses of particular genotypes to environmental characteristics is another important area to which crop simulation models have made a contribution. Such an application was reported by Field and Hunt (1974) to help determine the optimum response of lucerne growth to temperature in eastern Canada. Lower production in the latter part of the season was thought to be due to increased ambient temperatures, although it was difficult to confirm this experimentally due to the confounding influences of various combinations of day, night and soil temperatures. Using what was known of basic temperature responses from controlled environment experiments, the authors constructed a model to calculate the degree to which seasonal changes in temperature controlled lucerne growth. The results supported the hypothesis, and led to the suggestion that breeding work should be directed at selecting clones with more uniform performance at different temperatures. This was subsequently explored in actual breeding work by McLaughlin and Christie (1980).

A development of this approach has been to use long sequences of historical weather data and crop models to test the likely performance of a 'novel' genotype in a target environment. Differences in predicted yields from year to year give an estimate of the likely risk faced by a farmer in choosing to grow that genotype. This approach is particularly useful in variable environments such as, for example, the semiarid tropics, which are characterized by variability in the amount and temporal distribution of rainfall. These areas pose special problems for effective selection of improved genotypes, as the relative importance of different growth processes in determining final yield, and consequently the value of different traits, may differ between environments and between years in the same environment. It is expensive, if not impractical, to assess the value of different plant types using conventional multi-site, multi-season trials. This in turn, restricts the amount of information available to evaluate the risks associated with different plant traits that farmers are likely to face over longer periods of time. As an example, Bailey and Boisvert (1989) used a crop model coupled to long-term weather data to evaluate the performance of a range of cultivars at several locations in the semiarid areas of India by incorporating economic concepts of risk efficiency. They found that the ranking of the cultivars differed from that obtained with the traditional Finlay and Wilkinson (1963) approach, and depended crucially on the simulation of yields, and therefore on the ability of the model to accurately simulate the crop's response to water deficits.

Using a similar approach, Muchow *et al.* (1991) explored the consequences of maize and sorghum breeders selecting for: (i) greater rate of soil water extraction by the root system and (ii) a higher WUE. Their simulations showed that in the first case, the resulting faster exhaustion of soil water supply early in growth led to a 20–25% likelihood of yield loss due to lack of rain later to recharge the profile. In the second case, there was a yield gain in all years if the higher WUE was associated with no change (or increase) in radiation use efficiency (RUE), but a 30% chance of yield loss if the increased WUE was associated with lower RUE.

In a subsequent study, Muchow and Carberry (1993) used models for maize, sorghum and kenaf based on the CERES crop models to analyse three crop improvement strategies – modified phenology, improved yield potential and enhanced drought resistance. They found that there was no clear yield advantage of the traits in all years, and that the choice of plant type would depend on the farmer's attitude to risk. They defined a subset of cultivars as 'risk-efficient', characterized by a higher mean yield or lower standard deviation. However, the problem remained of how feasible it is in practice to modify the plant in the way shown by the simulations – a higher transpiration efficiency (g dry matter (DM) (g H_2O) $^{-1}$) was shown to be beneficial, but this is generally a very conservative parameter with little genetic variation. The work highlighted clearly the dangers inherent in using conventional selection techniques alone – traits selected for superior yields in a few years only could be very unrepresentative of their performance over a much longer time span – but also underscored the need to temper simulation results with information from field experimentation as to what was realistically achievable.

In rice, Aggarwal *et al.* (1996, 1997) used the ORYZA1 model for investigating effects on grain yield of various traits such as developmental rates during juvenile and grain-filling periods, leaf area growth, leaf N content, shoot/root ratio, leaf/stem ratio, and 1000-grain weight. Because of the lack of feed-backs built into this model, however, changing any one of these parameters generally changed yields in the expected way, with the exception of the phenological parameters, which interacted with year-to-year variability in weather. However, these changes were generally small, and they concluded that all parameters need to be increased simultaneously if there is to be any increase in yields – increasing one parameter alone has little effect. They also made the point that increased N applications might be necessary to express the effects of genotypes with higher yield potential as current N practices may be masking this potential. Yin *et al.* (1997) also used the ORYZA1 model to investigate the effect of variation in pre-flowering duration on rice yields at IRRI in the Philippines, at Hangzhou in China and at Kyoto in Japan. They concluded that the pre-flowering duration was about right for most modern cultivars – if it was any shorter, yield would be sacrificed, any longer and the number of cropping seasons possible per year would be sacrificed. In another study in West Africa, Dingkuhn *et al.* (1997) used the same ORYZA1 model to investigate traits that would enhance the competitive ability of rice against weeds (see Section 4.7 for further details).

Similar approaches have been used to assess the effects of different phenology in different varieties on grain yield for sorghum (e.g. Jordan *et al.*, 1983; Muchow *et al.*, 1991), rice (O'Toole and Jones, 1987) and wheat (Stapper and Harris, 1989; Aggarwal, 1991). Hammer and Vanderlip (1989) simulated the impact of differences in phenology and radiation use efficiency on grain yield of old and new sorghum cultivars. Jagtap *et al.* (1999) used the CERES-MAIZE model to show that short duration varieties performed better than long-duration varieties at three sites in Nigeria, but that the risk of crop failure was high if N was not applied. Other modellers have used simulation analysis to design improved plant types for specific environments (Dingkuhn *et al.*, 1991; Hunt, 1993; Muchow and Carberry, 1993).

Although much interest was generated in this approach, at the practical level, few plant breeding programmes have adopted it in any significant way. Donald (1968) himself recognized several of the difficulties of the approach, dividing them into conceptual (i.e. whether the approach was valid – is there such a thing as a 'best' type?), and practical (i.e. could the approach be implemented – e.g. which selectable characteristics determined the 'best' type?). Of the latter, one of the most serious is the frequent lack of genetic variability in reality of the characters in question. For example, BEANGRO predicts increases in yields with an increase in days to maturity in the absence of temperature or water deficit, but it has been difficult to breed lines that mature later than existing cultivars (White, 1998). Similarly, most models predict that increasing photosynthesis rates will increase yields, but little success has been achieved in practice so far in selecting for genotypes with increased photosynthetic rates. A second major problem is that characters are often negatively correlated, so that selecting to optimize one results in a suboptimization of another (e.g. Kramer *et al.*, 1982), cancelling out any improvement. Models are not usually able to predict these negative correlations in advance, although Boote and Tollenaar (1994) considered possible compensation between traits such as photosynthesis rate and specific leaf weight, and concluded that there was little potential for selecting for higher photosynthetic rates. Nevertheless, even as recently as 1991, Rasmusson (1991) argued that ideotype design for a particular environment was a useful exercise for a plant breeder, as it helped to focus his/her attention on what was or was not known about the environment, what particular characteristics it was practical to select for, and promoted goal setting for particular traits.

However, for many traits, crop models may not yet be sophisticated enough to capture the subtle differences between genotypes. For example, Yin *et al.* (2000) explored the ability of the SYP-BL crop model to explain yield differences between genotypes in a recombinant inbred line (RIL) population of barley, in which a dwarf gene (*denso*) was segregating. When all input parameters were calibrated using data from one of the seasons, the model could explain only 26–38% of the yield variation between the genotypes in a second season. Apparently, this was partly caused by some of the variation being due to plant N status, which the model did not account for. However, when the model was calibrated with values from the first season for only three of the parameters, lodging score, pre-flowering duration, and fraction of biomass partitioned to the spike, and with the other model input parameters held constant at their across-genotype means, the model could explain 65% of the yield variation in the second season. The authors make the point that part of the relatively poor performance of the model may be due to the so-called 'genotype parameters' it uses actually varying across environments, and give an example showing that the post-flowering duration and specific leaf area parameters varied with plant N status. This limitation obviously depends on the model being used. They also observe that in most crop models yield is determined by the availability of assimilate (i.e. source-limited) rather than the availability of sites to receive the assimilate (sink-limited). Again, the relative importance of source and sink approaches will depend on the type of crop – Matthews and Stephens (1998a) use a sink-limited approach to model the yields of tea, as previous work had shown that yields were poorly correlated with the source strength (Squire, 1985).

3.2 Environmental Characterization

The aim of any plant breeding programme is to develop improved geno-types for a pre-defined target population of environments. The target population could be defined either geographically (e.g. wheat varieties for the Punjab region in India) or in terms of a type of environment (e.g. rain-fed rice cultivation). However, because the definitions of the target environments are generally rather broad, there is usually a range of different individual environments within any defined population. Traditionally, genotypes are evaluated using multi-environment trials (METs) to evaluate the performance of a genotype in a sample of environments from the target population of environments. However, in many METs, there is no measurement of how well the sample environments match the target population of environments. Progress, therefore, is often slow because of the need to sample sufficient environments over sufficient years to be sure that any gain from selection is real.

What is needed is a way of characterizing all of the different environments within the target population with an index, so that similar environments with the same index can be grouped together. Representative locations within environments with the same index could then be chosen, with trials established at these locations to evaluate the selected genotypes. Results obtained from trials at these 'benchmark sites' can then be extrapolated with some degree of confidence to other similar environments with the same index. The number of METs that need to be carried out, therefore, could be greatly reduced.

The question then arises as to which indices are the most appropriate to characterize environments. Angus (1991) has reviewed the evolution of approaches to describe climatic variability, ranging from agroclimatological indices, simple water balances, through to crop simulation models. These methods vary in their input data requirements and in their complexity. Cooper and Fox (1996) distinguished between *direct characterization*, or characterization based on the measurement of environmental variables such as water availability or the physical or nutrient status of the soil, and *indirect characterization*, based on measurement or estimation of plant responses in a particular environment. As plant breeders are interested in the way a particular genotype performs in different environments, indirect characterization, taking into account how plants perceive their environment, perhaps gives a more realistic index. One way of doing this that has had some success is to use probe genotypes, where a specific set of genotypes is selected based on their known reaction to an environmental factor encountered in the target population of environments (e.g. Cooper and Fox, 1996). The relative performance of the genotypes which comprise the probe set can then be used to judge the incidence of the environmental factor in METs. A second approach currently being explored is to use crop simulation models to predict how a genotype with a particular set of characteristics will perform under different environments, and to characterize these environments in terms of that genotype's performance. The model used in this way acts as a means of transforming raw meteorological data into a form that represents the way a plant perceives its environment rather than just a purely physical description.

As a comparison of the different approaches to environmental characterization, Muchow *et al.* (1996) calculated three indices of water deficit to characterize target environments at two locations in Australia for grain sorghum. The simplest index was based on rainfall and potential evapotranspiration only, but poorly characterized the two environments. The second, a soil water deficit index based on a soil water balance and variable crop factor, and third, a relative transpiration index calculated using a sorghum simulation model, were both successful in identifying groups of seasons having distinct patterns. However, groupings based on the relative transpiration index from the crop model accounted for a higher proportion of the annual yield variation.

Chapman *et al.* (2000b) developed this approach further to use the environment types (ETs) to replace the location × years (L×Y) interaction term in the analysis of variance of trial results. Using data from 18 locations over 17 years, they used the sorghum model to generate drought stress patterns that were then grouped using pattern analysis into three environment types: (i) low stress; (ii) severe end-of-season stress; and (iii) medium end-of-season stress. They found that these ETs had more consistent relationships with simulated yields than did categorization of locations and years by descriptors such as rainfall and latitude. The implication of these results for plant breeding programmes in the region was that random sampling of environments (the current approach) is unlikely to be the most efficient way of improving broad adaptation, and that selection of locations representing the three ETs would improve this efficiency. In a companion paper (Chapman *et al.*, 2000a), the same authors argue that weighting genotype performance by the relative proportions of the three ETs across all sites and all years would improve the precision of the broad adaptation value. Their results indicated that if simple averaging of yields to select genotypes had been employed over the last 80 years, hybrids with adaptation to a higher frequency of drought environments than the long-term average would have been developed.

In another example of this approach, Chapman and Bareto (1996) used a simple model to define the extent of adaptation environments for maize in Central America using phenology and drought tolerance as traits. Monthly minimum and maximum temperature data from 364 base stations in the region were interpolated spatially in a geographical information system (GIS), and then used to develop maps of flowering date and thermal time accumulated up to 70 days after sowing (DAS).

A major limitation to the use of crop models to characterize environments in this way, especially in developing countries, is the lack of input data both in spatial and temporal dimensions. This may be because of either poor spatial coverage (i.e. few stations with reliable long-term records) or due to the availability of only monthly mean data rather than the daily data required by the models. Interpolation methods within a GIS such as those used by Chapman and Bareto (1996) go some way towards addressing this problem, although the reliability of data between weather stations is often dependent on the method of interpolation used. However, the availability of agrometeorological data suitable for use with crop models in developing countries is improving gradually all the time, and may not be such a limitation in the future.

A second limitation to most of the above approaches for environmental characterization is the failure to take into account socio-economic aspects. Because crop production always takes place in a sociological context, attempts to change cropping practices or recommend certain types of genotypes used by farmers within the target environment may fail if this is ignored (e.g. Fujisaka, 1993). Information on the preference of farmers for such things as plant type, grain/stover ratios and quality of grain for cooking and eating would help to ensure that the goals of breeding programmes were consistent with farmer requirements. If possible, information of this nature needs to be represented spatially in a GIS and overlaid onto the biophysical characterization information. Crop simulation models could be used to predict plant type and grain/stover ratios, but most models to date do not incorporate aspects of quality such as taste or cooking characteristics.

3.3 G × E Interactions

As mentioned in the previous section, the traditional way of evaluating genotypes is through large numbers of METs. However, METs are generally conducted for only a few years, and are unlikely to sample the full range of seasonal variability at a specific location, particularly where temporal variation is high such as at many locations in the semiarid tropics, for example. Current approaches used by plant breeders involve partitioning the variation observed in such METs for a desired trait into that due to genotype (G), environment (i.e. location and season, E) and the interaction between genotype and environment (G×E). The G×E interaction term is then often treated as a source of error or bias in the analysis of genotypic variation, which has resulted in the theoretical framework on which selection methods have developed being biased towards broad adaptation (Cooper and Hammer, 1996). Where the G×E term is large, however, the usefulness of using genotype means across the sample environments as an index for selecting superior genotypes is reduced.

Recognizing that mean yields across all of the sample environments may hide important differences in response, a number of statistical methods have been developed to analyse G×E interactions. One approach is to characterize each sample environment by the mean yield of all genotypes grown in the trial at that site, and then use this mean as an index of productivity of the site (Finlay and Wilkinson, 1963). Yields of individual genotypes across all the environments are then regressed against their corresponding site indices, and the slope of the line is taken as the stability or responsiveness of the genotype. However, the approach is often criticized as the site index violates the assumptions of statistical independence, and the response of genotype performance is assumed to be linearly related to the site index, which may not always be the case. Moreover, it is difficult to relate the site index to specific environmental factors such as water deficit or temperature stress. New statistical tools to address some of these problems, notably the assumption of linearity between genotype performance and site index, are being developed to discriminate between genotypes and to explain G×E interactions (DeLacy et al., 1996). These include Additive Main effects and Multiplicative Interaction (AMMI) models, and pattern analysis.

However, the problem remains of the time and cost of running sufficient METs to generate the data needed for the analysis. Aggarwal et al. (1996) proposed a strategy for increasing the efficiency of this process, by using limited MET data to estimate genotype interaction scores by AMMI analysis for all test genotypes on one hand, and to identify groups of genotypes with similar interactions via pattern analysis on the other. Representative genotypes for each group could then be identified and their performance simulated with a crop model over a wider range of target environments. The interaction scores for these new environments are then estimated from the simulated responses and combined with the genotype scores from the original MET to extrapolate G×E interaction effects over the wider range of environments. They used the ORYZA1 rice simulation model to simulate the performance of 26 hypothetical genotypes, which were 'created' by random combinations of eight model parameters (leaf N content, fraction of stem reserves, leaf/stem ratio, relative growth rate of leaf area, specific leaf area, spikelet growth factor, and crop development rates before and after anthesis), in ten different environments. These were then grouped into six genotype groups (Fig. 3.1), from each of which one genotype was arbitrarily selected as the reference genotype representing that group. Data for eight new environments were then generated using the same 26 genotypes. A highly significant positive correlation was obtained between the estimated and simulated interaction effects for the new sites, indicating the potential for this type of combination of statistical analysis and crop modelling to extend the range of G×E interaction information.

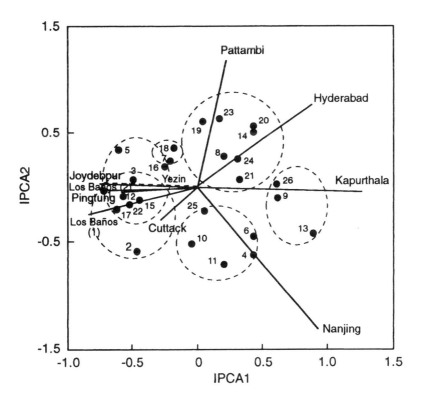

Fig. 3.1. Interaction bi-plot for the AMMI2 analysis of simulated G×E interaction data for 26 hypothetical genotypes simulated over ten environments. Lines join the environmental scores to the origin. IPCA scores are the multiplicative interaction scores for genotypes in the AMMI2 model, boundaries encircle groups of genotypes with similar interaction (after Aggarwal *et al.*, 1996).

A similar approach for sorghum in Australia was used by Hammer *et al.* (1996a), who investigated the effects of phenology, stay-green, transpiration efficiency, and tillering traits. As in the study by Aggarwal *et al.* (1996), they found that the partitioning of total variation between genotype, environment, and G×E interactions produced by simulation (4, 75 and 15%, respectively) were similar to that observed in the field, and that most of the G×E interaction variability was due to duration to maturity. In another study, Palanisamy *et al.* (1993) used a model based on the SUCROS family of models to predict the ranking of 11 genotypes in variety trials at three locations in India over 4 years. They successfully predicted the rankings of two of the top three genotypes, but concluded that the failure to do so with the other genotypes highlighted the need for further refinement of the methodology.

In another study, Acosta-Gallegos and White (1995) used the BEANGRO model to examine the length of the growing season at three sites in the Mexican highlands for 10–18 years. For two sites, long growth seasons and an early onset of the season were associated with greater probability of adequate rainfall. At the other site, total rainfall was lower, and was uncorrelated with the onset or length of the season. They proposed two types of cultivars – one with a growth cycle that becomes longer with early plantings for the first two sites, and a cultivar with a constant, short cycle for the third site. Similarly, Bidinger *et al.* (1996) used a simple crop model (Sinclair, 1986) to analyse G×E interactions for pearl millet in terms of differences among genotypes in the capture of resources, the efficiency of their use, the pattern of partitioning to economic yield, and their drought resistance.

Another way in which crop simulation models may be able to contribute in the area of G×E interactions is to reduce the amount of unexplained variability in the G×E term in the analysis of variance (ANOVA) of trial results. It has been recognized for some time that variation due to G×E interactions is amenable to selection if the environmental basis could be understood (e.g. Comstock and Moll, 1963). This has led to the concept of repeatable and non-repeatable G×E interactions (Baker, 1988), the repeatable part of which could be used as a basis for selection for specific environments. Crop models offer a way of predicting quantitatively the repeatable portion of this variability. A number of studies have used crop models to understand G×E interactions for various characteristics, but have not yet, to our knowledge, been used to partition the G×E interaction variance term to allow greater sensitivity of the analysis of variance. Much of the work to date has focused on G×E interactions in relation to phenology, probably because this is one characteristic in which the natural variation is greater than the resolution of most models. For example, Muchow *et al.* (1991) used a model to show that it was better to use a longer-maturing variety of sorghum than the standard cultivar (Dekalb E57+) at one location in Australia, while at another location, there was a 50% chance of yield loss from using either a shorter- or longer-maturing variety compared to the standard. An important point was that there is no clear advantage in all years of selecting a particular cultivar type. They were able to determine the probabilities of particular outcomes by using long-term weather with the crop model, which would have been time consuming and costly by a traditional MET approach.

Following on from the studies just described, some interesting work is emerging from the Agricultural Production Systems Research Unit (APSRU) group on linking crop simulation models to models of plant breeding systems as a way of understanding how the efficiency of plant breeding as a search strategy in a particular 'gene-environment landscape' could be improved. Chapman *et al.* (2002) describe an approach demonstrating how the flow of genes through breeding programmes can be investigated by incorporating assumptions about the links between individual alleles of genes and their corresponding phenotypic characteristics into a crop simulation model. In their study, four phenotypic traits were investigated – transpiration efficiency, flowering time, osmotic adjustment and stay-green. Each trait was assumed to be controlled by two genes, each with two alleles, giving five evenly distributed levels of expression depending on the number of alleles with positive effects present in the genotype. The APSIM-SORG model was then used to predict the yield of a crop with a particular genetic complement in a variable set of dry-land environments, grouped into three patterns of drought stress. Yields resulting from every possible combination of alleles were simulated in a large range of different environments (including different years and locations), and the results used as inputs into the QU-GENE model, which simulates different plant breeding systems (Cooper *et al.*, 1999). They found that different plant breeding strategies resulted in the accumulation of favourable alleles at different rates (Fig. 3.2), and, interestingly, that complex epistasis (gene–gene interactions) and G×E interactions emerged at the crop level, despite the assumption that the effects of different genes were simply additive. For example, alleles for the stay-green characteristic were not fixed until those for early maturity had first been fixed. The order of selection of different traits is, therefore, important.

Yin *et al.* (1999a) describe initial efforts to use crop models to improve the accuracy of analysis of quantitative trait loci (QTL). To identify QTL for SLA in 94 recombinant lines of barley, measurements of SLA were made at six times during the season (five of these were on the same date for all the lines, while one was at flowering, and hence differed between lines). Based on these measured data, between one and three QTL for SLA were found for all the sampling dates, and a dwarfing gene (*denso*) was found to strongly affect SLA. However, when a simple model based only on temperature was used to rescale the SLA measurements for direct comparison at the same development stage rather than chronological age, fewer QTL for SLA were found, and the presence of the *denso* gene did not affect SLA.

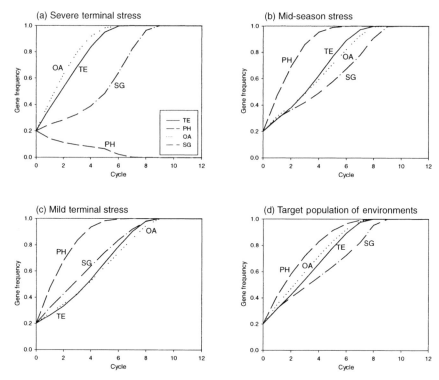

Fig. 3.2. Mean changes in the gene frequency for + alleles associated with four physiological traits: transpiration efficiency (TE, average of five genes), phenology (Ph, three genes), osmotic adjustment (OA, two genes) and stay-green (SG, five genes) given four different selection environments (from Chapman *et al.*, 2002).

The correlation in the first instance was found to be due to the gene's effect on the duration of the pre-flowering period, rather than on SLA directly. The authors suggest that the use of crop models in this type of analysis could be useful when investigating traits with a dynamic nature, such as leaf N content, biomass partitioning fractions, WUE and RUE.

An interesting suggestion is made by Hammer *et al.* (1996a) that another way in which models could help is to explore the interaction between management practice and genotype for different target environments, something that is not often accommodated in crop improvement programmes. They make the point that this interaction can be as important in assessing the value of a genotype as interaction with the physical environment. To our knowledge, there are no examples of studies investigating this aspect.

As already mentioned in Section 3.1, a major limitation of using current crop models in accounting for G×E interactions is the resolution and accuracy of the model in comparison to the subtle differences between genotypes commonly observed in many well-conducted multi-environment trials. For yield, these differences may be in the order of 500 kg ha^{-1} or less, which is probably less than, or at least near to, the resolution of most crop models. This level of resolution is due both to uncertainties in the input data used by the model (Aggarwal, 1995), and to inaccuracies introduced by the structure of the model itself.

It is this last factor that poses a dilemma in the application of crop models to crop improvement programmes – the simulated predictions are an inevitable consequence of the assumptions made in modelling the trait; by their very definition models are simplifications of a complex reality. However, for models to be able to capture the small differences between genotypes, they must be sufficiently detailed to simulate the interactions of growth and development in a particular environment. The dilemma is what constitutes 'sufficient detail'. One school of thought (e.g. Loomis, 1993) argues that more detailed models are required that are capable of simulating processes approaching the gene level. Some attempts have been made in this direction (e.g. Hoogenboom *et al.*, 1997; Yin *et al.*, 1999a, 2000). The criticism already discussed in Section 3.1 that most models do not adequately account for physiological linkages between traits (Lawn, 1994) would also support the argument for greater model sophistication. Often models do not incorporate these linkages because we do not have the knowledge of how they operate, and if they are included, their description is usually empirical rather than mechanistic (Mutsaers and Wang, 1999).

A contrasting point of view is that simpler crop physiological frameworks that are more readily aligned with plant breeders' modes of action are required (e.g. Shorter *et al.*, 1991). Hammer and Vanderlip (1989) were able to capture genotypic differences in RUE and phenology with a simple model, but such studies where simulation analyses of variation in a trait have been confirmed in the field are rare. Certainly, it seems logical that if crop models are to be incorporated into a crop improvement programme, it is essential that the parameters are easily and simply obtained, so that breeders can use them and apply them without substantial investment in time and data collection. Cooper and Hammer (1996) argue that crop physiologists have not generally appreciated this constraint faced by breeders, and have therefore not been able to adequately extend their often very relevant findings to 'real life' breeding programmes. However, it remains to be seen whether it is possible to resolve this dilemma of whether models of sufficient detail to discriminate between genotypes, yet requiring only limited input data, can be developed. The two approaches may not necessarily be mutually exclusive – Shorter *et al.* (1991) have suggested that the best way forward is to take a simple framework as the starting point, and add additional detail as necessary to describe the traits the plant breeder is interested in. A danger of this approach, which needs to be guarded against, is that the resulting model may reflect the prejudices of the user, and contain only the components that he/she thinks are important.

The other major limitation with current models is that not all of the traits that plant breeders are interested in are accounted for by the models (Hunt, 1993). Most crop models are designed to predict crop yield, but few crop improvement programmes focus on this characteristic only. Pest resistance and harvest quality, for example, are often of equal, if not greater, importance, but are not generally included in crop models (White, 1998). Some attempts to take these characteristics into account have been made – Piper *et al.* (1993), for example, used the SOYGRO model to explore the influence of temperature on oil and protein content in soybean. Similarly, recent advances in coupling pest models to crop models (e.g. Batchelor *et al.*, 1993) should make it easier to assess the effects of pest damage on crops, although further development is obviously needed to take into account complex mechanisms of disease resistance such as increased lignification, or changes in tissue N content.

While crop models have the potential to make an important contribution to the crop improvement process, Hammer *et al.* (1996a) warn that there are many issues faced by plant breeders where modelling may be of limited value. Issues associated with pests and diseases and some soil physical and chemical factors cannot be readily incorporated into existing models owing to lack of knowledge on the complexity of interactions with the crop. They suggest that, in most cases, such issues are best dealt with in other ways.

Crop Management

4

Robin Matthews

Institute of Water and Environment, Cranfield University, Silsoe, Bedfordshire MK45 4DT, UK

As crop–soil simulation models are designed to predict crop-level responses, it is perhaps not surprising that a large proportion of the work described in the literature in which such models are used is in relation to various management options of a single crop. Much of this modelling work has focused on understanding the interactions between the various factors influencing crop growth and development, such as water and nutrient supply, biotic stresses, and the timing of planting and harvesting of the crop in relation to the prevailing environment. This has led on to using the models to find optimum management practices for these factors in particular environments, generally with the purpose of maximizing yields. In this chapter, we look at examples where models have been used in this way.

4.1 Yield Gap Analysis

Before any improvements to crop management practices are made, it is useful to know what the potential yield[1] of the crop is in the region of interest, how large the gap is between this potential yield and yields actually being obtained, and what factors are causing any discrepancy between potential and actual yields. Pinnschmidt *et al.* (1997) define yield gap as the difference between an attainable yield level and the actual yield. It is affected by various constraints and limitations, such as cultivar character-

---

[1] Here we define potential yield as that yield determined by solar radiation, temperature, photoperiod, atmospheric CO_2 concentration and genotype characteristics only. Water, nutrients, and pests and diseases are all assumed to be non-limiting.

istics, cropping practices, weather and soil conditions, and stresses due to pests, diseases and inadequate water supply. An analysis of the yield gap allows a quantification of the likely benefits to be gained by embarking on a programme to improve crop management, and identification of the factors that it is worthwhile concentrating research resources on. Crop models offer a way of estimating what the potential yield of a crop is, and a step-wise analysis of the various inputs can help identify the limiting factors.

An example of such an application is provided by studies conducted at different sites in India, during evaluation of the groundnut model PNUTGRO (Boote *et al.*, 1991; Singh *et al.*, 1994). Using parameters for the standard cultivar Robut 33-1 (=Kadiri 3), the model predicted that potential yield as determined by climatic factors alone was achieved at about one-third of the sites, but at many locations poor growth and low yields could not be attributed to weather conditions. It was concluded that other factors such as soil fertility and pests were causing a yield gap at some sites. Subsequent research focused on these problems.

In a similar study in wheat, the WTGROWS model was used to predict potential wheat yields across India (Aggarwal and Kalra, 1994; Aggarwal *et al.*, 1995), which were compared with the economic optimum yield and actual yields across a range of latitudes (Fig. 4.1). Results showed that yields increased with increasing latitude and at more inland sites, primarily because of variation in temperature. Average actual yields were less than 60% of potential yield – although actual wheat yields had increased considerably over the preceding 25 years to 3000 kg ha^{-1}, they concluded that the yield gap was still at least 2000 kg ha^{-1}. Further analysis suggested that about 35–40% of this gap was due to delayed sowing – most farmers are sowing later than the optimal planting date as rice/wheat systems are becoming more common. Rice matures in October/November, which is the optimal wheat planting date, and as rice is more profitable, farmers try to maximize its yield, which underlines the importance of taking the whole system into account in any analysis. Irrigation inefficiencies and variability in fertilizer use were other important factors limiting wheat yields. There is no evidence, however, that the findings of this work have been used by planners or to prioritize research, although it should be noted that WTGROWS itself is currently being used for yield forecasting (P.K. Aggarwal, New Delhi, 2000, personal communication).

In another example, Pinnschmidt *et al.* (1997) collected data on crop and pest management practices, soil conditions, weather, crop performance, and biotic and abiotic stresses from 600 plots in farmers' rice-fields in the Philippines, Thailand and Vietnam. The CERES-RICE model was used to estimate potential and N-limited attainable yields, and a simple empirical approach was used to estimate yield trends based on fertilizer N and soil organic matter. The gaps between these predicted attainable yields and actual yields ranged from 35 to 55% in the different countries. In Thailand, it was shown that much of this was due to N limitations, resulting from

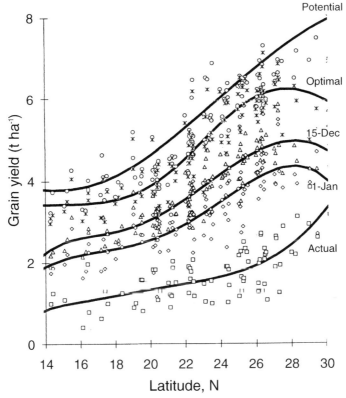

Fig. 4.1. Potential, economic optimum and actual grain yield of wheat as a function of latitude. Also shown are the simulated yields on 15 December and 1 January sowings to illustrate the contribution of later sowing to the yield gap (from Aggarwal *et al.*, 1995).

low soil organic matter and low fertilizer inputs. Other factors such as pests and disease damage and water stress were important in the Philippines and Vietnam. This type of information can help in setting priorities in studying and managing yield-limiting factors, although again there is no evidence to date that it has been taken up and used by anyone (H. Pinnschmidt, Denmark, 2000, personal communication).

Van Ranst and Vanmechelen (1995) developed three simple crop models to estimate potential yields, water-limited yields and yields limited by soil suitability, for the north-west region of the Cameroon. As a demonstration of the approach, these models were used within a geographical information system (GIS) framework, and maps were produced of the predicted yields at the three different production levels. However, the authors make the point that the lack of accurate environmental data for operation and validation of the crop models is a serious constraint that must be given urgent priority.

In Mali, public investment in irrigation schemes to try to capitalize on the expected high potential yields of rice due to high solar radiation and adequate water have not been as successful as hoped. In an attempt to identify why the expected yields were not being obtained, Dingkuhn and Sow (1997) used the ORYZA_S rice model to study the spatial, seasonal and year-to-year variability of potential rice yields in the region as a function of planting date. Results indicated that potential yields are primarily driven by temperature, and that the major physiological determinants of yield were: (i) crop duration, which is very variable due to flood-water temperature; (ii) leaf area expansion, which is susceptible to chilling; and (iii) spikelet sterility due to heat or chilling. Yields varied from 4 to 10 t ha^{-1}. The results were used to propose environment-specific research foci.

In a similar example, van Keulen (1975) was able to show, through simulation studies of growth in semiarid conditions in Israel and the Sahel, that in many years production was limited by nutrient deficiency rather than lack of water, as had been commonly thought. This insight led the way for subsequent comprehensive research projects on primary production in both of these regions. In Zambia, Wolf et al. (1989) used a model to simulate identification of the factors limiting maize yields for the main land-units, and found that rainfall was limiting only if it was less than 800 mm. At higher levels of rainfall, the main constraint to higher yields was nutrients, indicating that there would be a response to fertilizer. It is not known if the results from this work had any impact.

An interesting use of a crop model to evaluate possible causes for change in crop yields over time in a given region is provided by Bell and Fischer (1994). Farmers' yields of wheat in a region of Mexico had increased by nearly 60 kg ha^{-1} year^{-1} between 1978 and 1990 due to improved varieties, crop management and weather variation. The CERES-WHEAT model was used to predict potential yields in the region assuming no change in cultivar or management over the time period. The analysis showed that yields would have declined over this period because of increased temperatures, and that the true yield gain, attributed to improvements in genotype and crop management, was in fact 103 kg ha^{-1} year^{-1}. However, despite these gains, average farmers' yields, having risen 50–75% over the period in question, were still considerably lower than the potential yields predicted by the model, indicating that there is still scope for improvement.

The main impact of all of these studies has been to focus research activities on the major factors limiting yield, although in some cases there is no evidence that this information has been used. It is difficult to quantify in monetary terms the value of such work, as this depends on the outputs of the downstream research. Nevertheless, it would seem logical that using models to identify limiting factors and prioritize research effort in these areas is a more efficient way forward than carrying out large-scale field experiments and finding out afterwards that the wrong factors were being investigated. Ways of disseminating this information to the relevant researchers, however, need to be improved substantially. It is tempting to suggest that if a sound modelling study had been carried out in Mali *before* rather than *after* public money had been spent on irrigation schemes, the results might have been more productive than was the case.

4.2 Soil Surface Management

The condition the soil is in can have a major influence on the crop that is subsequently grown in it. It is, therefore, of interest to know what effect various soil management practices have on crop growth and yield. Freebairn *et al.* (1991) used the PERFECT model to simulate the effects of various management practices, such as crop/fallow sequences, tillage and addition of soil ameliorants, to modify different soil physical processes including infiltration, evaporation and erosion. They also used sequences of historical weather data to look at long-term decline (100+ years) in yields associated with soil erosion. Results showed that annual soil loss was much greater when previous crop stubble was removed.

In another study, Singh *et al.* (1999) used a soybean–chickpea sequencing model to extrapolate 2 years of experimental data investigating the effect of two land preparation techniques – broadbed-and-furrow (BBF) and flat – for two depths of soil in India. Using 22 years of historical weather data, the model simulations showed that in most years, BBF decreased runoff from the soil, but had a marginal effect on yields of soybean and chickpea, although these effects tended to be larger in dry years. The decreased runoff was associated with a concomitant increase in deep drainage from the BBF treatments.

There is no record for either of these studies of any practical impact they might have had.

4.3 Planting

In most environments, the time a crop is sown can have a major influence on its growth during the season, and therefore on its final performance. This is particularly the case in variable environments, or where there is a strong seasonal effect. In many tropical and subtropical regions, for example, planting decisions await the onset of a rainy season, and the available soil water reservoir is often only partially recharged over the dry season. In such cases, planting too early may result in poor establishment if the soil water status is insufficient, while planting too late may mean that the crop encounters drought stress towards the end of the season, the time in many crops when the economic yield is being determined.

For example, Omer *et al.* (1988) used a crop model and 11 years of climatic data to determine the optimum planting period in the dryland region of western Sudan, by generating probability distributions of a water-stress index resulting from different planting dates. The analysis showed a distinct optimum planting period of 20 June to 10 July, with planting in early July as the most likely for best production, which agreed well with general experience. In a similar study, Singels (1992, 1993) used the PUTU wheat growth model to determine optimal wheat-sowing strategies in South Africa using 50 years of historical weather data. Highest mean production was simulated when the entire available area was planted on the first possible date after 5 May. The starting date of the optimal sowing period identified by the simulations did not differ markedly from those recommended by the South African Department of Agricultural Development, although the last date of the optimal period occurred earlier than those recommended. The analysis indicated that profit-maximizing, risk-averse producers should delay sowing until 9 June and then plant the total available area as soon as favourable sowing conditions occur. A similar conclusion was reached by Williams *et al.* (1999) for grain sorghum in Kansas – extremely risk-averse managers would generally choose somewhat later sowing dates, earlier maturing hybrids and lower sowing rates than less risk-averse or risk-preferring managers. In Australia, Muchow *et al.* (1991) showed for sorghum that sowing later on a full-soil profile of water was always better than sowing earlier on a half-full profile.

Similarly, Singh *et al.* (1993) and Thornton *et al.* (1995b) describe work using the CERES-MAIZE model, calibrated for local field conditions in Malawi, to determine the optimum planting window and planting density for a number of varieties currently grown there. In northern India, Aggarwal and Kalra (1994) used the WTGROWS model to show that a delay in planting date decreased wheat yield, in part by subjecting the crop to warmer temperatures during grain filling. These results confirmed experimental data presented by Phadnawis and Saini (1992) for New Delhi. Hundal *et al.* (1999) used the CERES-RICE model to evaluate the age of seedlings at transplanting, number of seedlings per hill and plant population for rice growing in the Indian Punjab. Results showed that the optimum date of transplanting for rice was 15 June, but that earlier-transplanted (1 June) rice may perform better if seedling age is reduced from 40 to 30 days. Increasing plant population increased rice yields. Saseendran *et al.* (1998) also used CERES-RICE to determine the optimum transplanting date for rice in Kerala, southern India. Similarly, Hoogenboom *et al.* (2001) used CROPGRO-PEANUT to determine optimum planting date for groundnut in Andhra Pradesh, finding that later planting dates had a higher yield potential than earlier planting dates. Farmers, however, prefer to plant early to avoid pest and disease damage prevalent with late planting, which the model does not currently simulate.

In Bangladesh, Timsina *et al.* (2001) used the SWAGMAN Destiny model to evaluate the optimum time of planting of short-duration mung bean during the period between the main wheat and rice crops. During this time

(March–June), the rainfall is somewhat erratic, leading to either water deficits or waterlogging, to both of which mung bean is susceptible. They found that where soils were undrained, a March planting was best, whereas for drained soils, planting in April was optimum.

In Mozambique, Schouwenaars and Pelgrum (1990) used a crop model to simulate maize production over 28 years for different sowing strategies. They found that the maximum annual production depended almost completely on losses caused by pests and diseases and postharvest losses. However, if the criterion was to minimize periods of food shortage, the preferred sowing strategy depended on water availability.

In Australia, Clewett et al. (1991), while designing shallow-dam irrigation systems, considered two planting options – the first was to plant as soon as there was sufficient rain to ensure crop establishment, while the second was to delay planting until there was sufficient runoff to provide irrigation so that crop production could be assured. The first option was shown to have the higher long-term mean production, although this was accompanied by a much higher variability of production. Also in Australia, Muchow et al. (1994) assessed climatic risks relative to planting date decisions for sorghum growing in a range of soils in a subtropical rain-fed environment. Yield response was associated closely with differences in leaf area development and degree of depletion of the water resource brought about by differences in sowing date. It was suggested that decision makers could use the information taking into account their risk preferences, but no evidence is presented of this having happened.

A general approach to generating the information required to assist in making sowing decisions in climatically variable subtropical environments is presented by Hammer and Muchow (1992). The approach involved coupling a sorghum growth simulation model to long-term sequences of climatic data to provide probabilistic estimates of yield for the range of decision options, such as sowing date and cultivar maturity, for a range of soil conditions. The likely change in the amount of stored soil water with delay in sowing was also simulated to account for the decision option of waiting for a subsequent sowing opportunity. The approach was applied to three locations in subtropical Queensland, Australia. Production risk varied with location, time of sowing, soil water storage and cultivar phenology. The probabilistic estimates presented of yield and change in stored soil water could assist decision makers with risky choices at sowing in subtropical environments.

The density of planting is another characteristic that has been investigated with crop models. Much early work on determining optimum planting density used static models (e.g. Stickler and Wearden, 1965) which related plant population density to overall yield and to its components, such as yield per plant. More recently, crop models have been used to develop and confirm these relationships for particular environments. Keating et al. (1988), for example, used the CERES-MAIZE model to examine the effects of planting density on maize yield as influenced by water and N limitation in Kenya, and found that the density for maximum yield increased as N supply increased. Singh *et al.* (1993) carried out a similar analysis in Malawi.

Due to their dynamic nature, however, crop models offer a way of exploring variations in these relationships between environments or from year to year, and can therefore be used to quantify the risk faced by the farmer of choosing a particular planting density in a particular environment. Quantification of this risk is particularly important in variable environments, as a low planting density may mean that there is at least some yield in a poor year even though yield may be sacrificed in a good year. A high planting density, on the other hand, may mean that maximum yield is obtained in a good year, but that there is no yield at all in a poor year. The most appropriate strategy for planting density, therefore, will depend on both the specific environment and the farmer's attitude to risk – a risk-averse farmer should use a low density, whereas a more risk-tolerant farmer may opt for a higher density, and therefore maximize his/her income over the long-term despite total crop failures in some years.

A good example of this type of analysis is provided by Wade (1991), who used the SORKAM sorghum model to analyse risk associated with different planting densities (low, standard and high) at three contrasting sites in Australia. At one site (Katherine) it was always better to opt for high density and narrow rows. At the other two sites (Emerald and Dalby) standard practice appeared to be the best compromise – over 30 years, the high planting density gave higher yields in 5 years, but crop failures in 14 years, compared to the standard. Low planting density at these sites gave only four crop failures, but yields were lower in good years. The model was also used to investigate the effects of variation in stand uniformity – it was predicted that a poor distribution of plants gave 11% less yield than the same plant density but with uniform distribution, whereas variation in both plant density and plant distribution gave 25% less yield. It was suggested that this kind of analysis might be able to help in making decisions on whether it is necessary to replant a poorly established stand or not. In the Philippines, Lansigan *et al.* (1997) used the ORYZA_W model to generate probability distributions of rain-fed lowland rice yields for different planting densities and seedling ages at transplanting. Stochastic dominance analysis was then applied to identify risk-efficient management options in the light of farmers' attitudes towards risk.

The method of planting has also been studied with the aid of a crop model. Muchow *et al.* (1991), for example, found that, except at the highest yield levels, no-till sowing gave higher yields and reduced risk of negative gross margins compared with conventional planting methods. The choice of the appropriate crop to grow is another important decision that a farmer must make at planting time. In the same study, Muchow *et al.* (1991) also used the model to evaluate whether sorghum or maize was the best crop for a particular environment. Results showed that in the long term, maize was the most profitable, although in about 35% of the years, sorghum performed best. A similar study in the bimodal rainfall areas of Kenya showed that if the rains are early, then maize should be sown; if delayed, then sorghum or millet is better (Stewart, 1991).

A limitation of most of these models, particularly in studying effects of planting density in crops in subsistence agriculture, is that most assume a uniform canopy structure across a field. While this approach has shown itself to be successful in describing conditions in crops in mechanized agriculture in developed countries, it is woefully inadequate in reflecting reality in many farmers' fields in developing countries, where the planting method or poor stand establishment often results in single plants arranged more or less randomly. Such canopies, where leaf area tends to be clumped and a significant proportion of the projected area is bare ground, violate the implicit assumption made in most models that leaf area is distributed spherically. Rather than solar radiation penetrating the canopy only vertically, as the uniform canopy approach assumes, it can reach the plant from all sides as well. The models, therefore, probably significantly underestimate light interception and water use in these cases. To our knowledge, few crop models consider the microclimate of individual plants in a way that would realistically reflect crop canopies in farmers' fields.

It is difficult to assess whether any of this research has had any impact on changing farmer practice. In most cases, it appears that the studies merely confirm what was already known. Over the years, farmers have developed their own rules of thumb for determining the optimum planting date and densities of their crops, and in many cases, there is probably little scope to improve on these rules through the use of crop models and optimization techniques. However, in variable environments, or where new options are being considered, such as the introduction of new crops or cropping sequences (see Section 5.1 for an example of this), models can provide useful insights into when is the best time and what is the best way to plant.

4.4 Water Management

Aspects of water management have also been covered under other headings – for example, determination of planting date is often dependent on the availability of soil water (e.g. Muchow *et al.*, 1991). Similarly, many applications of crop models relating to water management have been incorporated into decision support systems for scheduling of irrigation, which are discussed in more detail in Chapters 9 and 10. The following are some examples of how crop models have been used as tools to aid research in water management.

Matthews and Stephens (1997) used the CUPPA-TEA model to provide initial evaluations of different irrigation options in terms of yields and profitability for tea growing in Tanzania. For mature, deep-rooted tea, the model predicted that yields start to decrease only when the actual soil-water deficit exceeds about 140 mm, and that irrigating more frequently than this does not produce any higher yields and may only add to production costs. The results also showed that water applied during the cool-dry season (June–August) at this site could be used by the crop as effectively as water applied later during the warm-dry season (September–December), provided there are no losses from runoff or drainage. Where irrigation dams have limited storage capacity, applying as much water as possible early in the season, up to the full water requirements of the crop, is the most effective way of using water. A preliminary economic analysis confirmed previous work that the profitability of irrigation is very sensitive to the price received for tea. When the price was higher than TSh900 kg^{-1} made tea (MT) (~ £1 kg^{-1}), irrigation was worthwhile, but there was no extra return from applying water when prices were lower. In another study, the tea model was used in a consultancy study to evaluate the feasibility of irrigation on a commercial estate in Zimbabwe.

Lansigan et al. (1997) used the ORYZA_W model to evaluate alternative management options for rice production in the light of farmers' attitude towards risk. The model was used to generate probability distributions of rain-fed lowland rice yields under different management scenarios which included water management (bund height, puddled soil depth, planting density and seedling age at transplanting). Stochastic dominance analysis was applied to identify risk-efficient management options.

In South Africa, du Pisani (1987) used the CERES-MAIZE model to help develop an index to characterize drought to give policy makers an objective measure to declare areas drought-stricken and to implement subsidy schemes fairly. The index used up until that time, the Palmer Drought Severity Index, had been found to be deficient in several aspects. The model was validated using yield data collected at a wide range of localities within South Africa with a mean annual rainfall of 550–870 mm and grain yield of 0.99–8.60 t ha^{-1}. It is not clear how much improvement over the previous index was obtained, or whether the revised index is now in use by policy makers. The model was also evaluated for its potential for forecasting crop yields using observed early season data combined with median data for the remainder of the season. Good agreement was found between observed and predicted end-of-season yields.

Stephens and Hess (1999) used the PARCH model to extrapolate field results of the beneficial effects of soil conservation techniques on yields over longer time periods than the few years of measurements. The results showed that runoff control and runoff harvesting produce significant yield increases in average years in both the long rains and the short rains. However, in dry years, there were only small yield increases in the short rains and negligible benefits in the long rains. In wet years, there were no significant yield increases due to water conservation in either season. The point is made by the authors that these simulations are a simplification of reality, and that in the real situation water conservation strategies may allow such adaptations as earlier planting or increased planting densities. It is not stated if this work was taken up and used in a practical way.

The PARCH model was developed further to investigate aspects of rain-water harvesting in Tanzania, resulting in the PARCHED-THIRST model (Young and Gowing, 1996). Rain-water harvesting is defined as the collection of runoff as sheet flow from a catchment area into an adjacent cropped area without storage other than in the cropped area. The purpose of the model was to: (i) design the most appropriate system given site characteristics by optimizing predicted crop yields; and (ii) act as a tool for technology transfer both from research to the farmer and from location to location. To facilitate the model's intended use in areas where few or no data are available, it represents the important hydrological processes using physical parameters that are readily available (e.g. soil survey) or can be easily measured or estimated. The identified target group was government and non-governmental organization (NGO) staff in extension and planning, and a number of workshops were held to train these in the use of the model. There is no record, however, of the degree of impact the use of the model has had on the development of rain-water harvesting in the region.

MacRobert and Savage (1998) describe the development of an interactive computer program, WIRROPT7, based on a modification of the CERES-WHEAT model that searches for the intra-seasonal irrigation regime that maximizes the total gross margin for a particular soil, weather and crop management combination, within the constraints of land and water availability. The use of the system is demonstrated for two soils near Harare in Zimbabwe, and for two strategies – maximizing gross margin per unit area with a frequent irrigation schedule, or maximizing the overall profits by reducing the application per unit area by irrigating less frequently, but growing a larger area. As the variability of yields and gross margins are higher with the second strategy, the farmer's attitude to risk will determine which strategy he/she adopts. The authors recommend that the model should be seen only as a guide to irrigation management due to its limitations such as lack of routines describing nutrients and pests, and the need to use mean historical weather data. The actual irrigation strategy should take into account real-time estimates of water use as well. It is not stated if the system is actually being used by farmers or extension staff.

A good example of the use of models to extrapolate results from a few year's experiments is given by Clewett *et al.* (1991), who were investigating the feasibility of storing ephemeral runoff in shallow dams for strategic irrigation of grain and forage crops. As these dams do not give a guaranteed supply of irrigation water, the productivity of shallow storage irrigation systems is still closely related to climatic variability. Ten years of experimental data were collected which indicated that such systems were viable (Fig. 4.2b). However, simulations over 60 years of historical weather data revealed that these results were biased – the experiments just happened to have been conducted in a 10-year period when the rainfall was above average (1968–1977 in Fig. 4.2a). Recommendations based on the experimental evidence alone would have been misleading in the longer term, and indeed even in the years immediately following the experimental work. This highlights the dangers of relying on experimental data alone, even from 10 years, which in many cases would be considered conclusive proof that a technology had been well validated. A combination of field work and simulation modelling in this case was shown to be essential in defining optimal agronomic practices in the semiarid tropics where rainfall is highly variable.

Limitations of water balance models depend on the model and the use to which it is to be put. The simplest assume that the soil can be represented by a single layer, and that crop transpiration is equivalent to potential evapotranspiration until a specific water deficit is reached, then it declines linearly. Such models do not explicitly take into account the distribution and effectiveness of the root system. More sophisticated models may divide the soil into a number of layers and assume a particular distribution of roots within these layers. Vertical water movement in some of these is assumed to be like a tipping bucket – an uppermost layer must be filled to field capacity before any movement of water to the next lower layer. In reality, water can begin movement to the next layer before a particular layer reaches field capacity. Most crop–soil models are one-dimensional in that they consider only fluxes of water in a vertical direction and not those in a horizontal direction. Lateral fluxes can represent significant quantities of water, particularly in sloping land, which characterizes much subsistence agriculture. Another limitation of most models is that they do not consid-

(a)   (b)

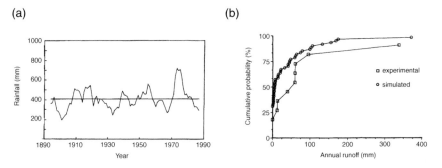

Fig. 4.2. (a) Five-year moving average at Richmond, Queensland. Long-term summer average rainfall is 404 mm. (b) Cumulative frequency distribution function for annual runoff from the gauged native Mitchell-grass pasture catchment at Richmond. Recorded data during both a 10-year experimental period 1968–1977 and a 60-year simulation period 1918–1978 are shown (from Clewett *et al.*, 1991).

er preferential flow of water down cracks in the soil, again a common occurrence in tropical soils.

4.5 Nutrient Management

In many tropical countries, fertilizer is a relatively expensive commodity. In Zambia, for example, structural readjustment of the economy in 1990 resulted in the removal of external donor-funded subsidies from fertilizer and other agricultural inputs. This had the effect of forcing many farmers to move from recently adopted continuous maize cultivation back to their traditional forms of cultivation such as *chitemene* and *fundikila* (Matthews *et al.*, 1992). Even those farmers who were able to continue growing maize needed to consider more carefully how to maximize their returns to a relatively more expensive input.

In such countries, the availability of N to the crop, and hence the efficiency of use of N fertilizers, is often highly variable, much of which is caused by variation in the prevailing climate and in soil conditions. Drought, for example, can restrict the uptake of N by the crop, as well as reducing rates of N mineralization. On the other hand, too high a rainfall can result in losses of N from the soil by leaching and denitrification. In highly alkaline soils, N can be lost by volatilization. Immobilization by microorganisms can also remove significant amounts of N temporarily from use by the crop. As all of these processes vary in importance according to the prevailing weather conditions, it is difficult, therefore, to define a single fertilizer strategy that is optimum in all seasons. As a result, there is often a mismatch between supply and demand of N, thereby reducing yields.

Field experiments conducted in environments with high climatic variability may give misleading results, as the years in which they are conducted may not represent the long-term average. In such cases, crop models provide a way of assessing the long-term risk of particular options, thereby complementing the experimental results. For example, Keating *et al.* (1991) used a modified version of the CERES-MAIZE model (CM-KEN) to investigate the factors influencing response to N in the Katumani region of Kenya, looking at variations in organic matter, mineral N and soil water at planting, runoff characteristics, plant density and timing of N applications. In all cases, the response to N varied according to the amount and pattern of rainfall in the season. Taking into account this year-to-year variability, they were able to use the model to plan a hypothetical development pathway involving the application of more N and increasing the density of planting. Similarly, in Australia, Verburg *et al.* (1996) used the APSIM-SWIM model to evaluate the effect over 33 years of different irrigation management strategies on the responses to N fertilizer application in sugarcane in terms of crop yield and nitrate leaching below the root zone.

Ways to optimize the timing and amount of applications of N to a crop have received considerable attention in developed countries, with the emergence of a wide range of decision support systems to provide recommendations to farmers (e.g. Falloon *et al.*, 1999), some of which are discussed in more detail in Chapter 9. Many of these N recommendation systems are based on the concept that additional N required by the crop is the difference between the total N required by the crop and the soil N supply. The crop N requirement is determined by the amount that is needed to achieve a specific yield, and the soil N supply by the properties of the soil, particularly the organic matter content and factors that affect the rate of microbial decomposition. The additional N requirement is usually adjusted by a fertilizer use efficiency to calculate the amount to be actually applied. However, year-to-year and site-to-site variability means that this approach requires the parameters to be recalibrated at each location over several years. As a way of understanding the causes of this variability, Bowen and Baethgen (1998) used the CERES model to explore systematically some of the factors influencing N dynamics in soils, taking into account crop N demand as affected by days to maturity, and the soil N supply as affected by the amount of soil organic matter (SOM), rainfall and initial soil mineral N. As might be expected, longer-maturing genotypes took up more N than earlier-maturing genotypes, and both high SOM levels and high initial soil mineral N levels resulted in higher crop N uptake. Interestingly, there was a maximum crop N uptake in relation to annual rainfall – at first, N uptake increased as rainfall increased the growth of the crop, but at rainfalls above 500 mm, crop uptake declined as leaching losses became more significant. It was shown how such results could be used to evaluate for different soils the trade-offs between the potential benefit of applying N fertilizer in terms of yield, and the environmental cost in terms of nitrate leached. A similar analysis was made by Alocilja and Ritchie (1993), who used the SIMOPT2:MAIZE model, based on CERES-MAIZE, to investigate the trade-off between trying to maximize profit and trying to minimize nitrate leaching.

Singh *et al.* (1993) used CERES-MAIZE to determine N response curves for two different maize cultivars and two different sites in Malawi over a number of years. They then used this data to calculate the economically optimum rate of N fertilizer application, although only the grain price was taken into account – despite the higher yields from hybrid varieties, farmers still prefer to grow the local landraces as they produce more stover which is used for fodder. A more accurate analysis would take into account the fodder/grain trade-off (R. Fawcett, University of Edinburgh, 2000, personal communication). Thornton *et al.* (1995b) took the analysis one step further by linking it to a GIS with spatial databases of soils and weather to analyse the influence of N management on crop yield and leaching potential at the regional level. Such a linkage not only allowed an analysis of the spatial variability due to different soil types and weather across the region, but also the temporal variation due to year-to-year differences in weather.

As part of the Simulation and Systems Analysis for Rice Production (SARP) project (described more fully in Section 12.4.2), ten Berge *et al.* (1997a, b) describe the development and testing of the ORYZA_0 model, a 'parsimonious' model based on incident solar radiation, bulk leaf N and a site calibration factor. The model was subsequently used by Thiyagarajan *et al.* (1997) to classify some irrigated rice soils in India into those in which the soil N supply was sufficient to meet crop demand up to first flowering, and those that required a basal dressing of N fertilizer. The results were then used to generate fertilizer N recommendation curves that identified different optimal timing of N application for the different soil N supply regimes. The model was also used in a similar exercise in China, where it was found that significantly higher yields were obtained by following the recommendations produced by the model compared to the local recommendations (ZhiMing *et al.*, 1997). Whether the model is able to make significant improvements over local recommendations in a wider range of environments remains to be seen. A limitation of ORYZA_0 is that it must be calibrated for each site, in this case by measuring the seasonal pattern of crop N uptake, which, because of year-to-year variation, makes it difficult to use the model in a predictive way. Also, no account is taken of the ability of the soil to act as a reservoir or 'bank' of N, i.e. N applied earlier in the season can remain in the soil for uptake by the crop later, so that it is not critical when the fertilizer is applied. The model only indicates how much N must have been applied by a specific crop age, and not when and how (i.e. as a single dressing or as split doses) it should be applied.

An interesting use of crop models is to provide a quantitative basis for response farming, a method of identifying and quantifying seasonal rainfall variability (Stewart, 1991). Traditional agricultural research has generally only considered fixed strategies of fertilizer management, but in reality farmers often make tactical adjustments to their management in the light of what they perceive as information relevant to the prospects for the forthcoming crop. For example, if a farmer perceives that the rainfall during a season is likely to be less than normal, he/she may decide to apply less fertilizer. As an example of analysing this approach, Keating *et al.* (1993) used the CERES-MAIZE model to examine the value of changing N-fertilizer application rates in line with predictions of the likely response to fertilizer based on the date of the season onset in the Machakos region of Kenya. The date of season onset had been found previously to be a useful predictor of the length of the growing season (e.g. McCown *et al.*, 1991), and hence it was thought that it might also be a predictor of the response to fertilizer inputs. However, results showed that a conditional strategy in which fertilizer application was adjusted in relation to onset date resulted in only a small increase in gross margins – applying some fertilizer was the highest priority as far as production was concerned. Nevertheless, risk analysis showed that the likelihood of negative gross margins was reduced by conditional application of fertilizer, suggesting that it may be an attractive strategy for more risk-averse farmers. Thornton *et al.* (1995a) carried out a similar analysis in Malawi by classifying seasons over a number of years according to their start (i.e. early, normal, late), and calculated maize yields for each group. Results showed that yields decreased the later the season started, and that the optimal rate of fertilizer application was 90 kg N ha^{-1} for early starting seasons, declining to only 30 kg N ha^{-1} for late starting seasons.

The effect of different strategies of N application on the environment is also becoming increasingly important. Singh and Thornton (1992) used the CERES-RICE model to investigate the effect of various nutrient management strategies on N leaching from rice fields in Thailand. Strategies considered were different levels of N fertilizer and green manure on two soil types (clay and sand), which were simulated using 25 years of weather data. Results showed that on a clay soil, leaching losses were similar for all treatments, but that there was a significant decline in rice yields when 50 kg N ha^{-1} urea or 4 t dry matter (DM) ha^{-1} were applied compared with 100 kg N ha^{-1} urea. This suggests that medium- to high-input agriculture on clay soils is not only productive, but it is also as environmentally sustainable as lower input agriculture. Leaching losses were considerably more on sandy soils than clay soils. Green manure applied at rates of 4 t ha^{-1} was shown to be able to substitute for 40 kg N ha^{-1} urea, but leaching losses from the green manure could be as high as from using inorganic fertilizers. In a similar study, Singh and Thornton (1992) used CERES-RICE to evaluate the effect of different fertilizer incorporation strategies on losses of N by ammonia volatilization from rice fields in the Philippines. Results showed that when urea was broadcast onto 5 cm of floodwater with no incorporation, N losses were 18 kg N ha^{-1} or 36% of that applied. The losses declined with increasing degree of incorporation, and were negligible when urea was deep-point placed.

A good example of how models can be used to help prioritize research in developing countries is given by McDonagh and Hillyer (2000), who developed a model describing N flows in crops to evaluate if soil N status could be improved through the use of legumes intercropped with pearl millet in northern Namibia. Data was collected on N fixation rates of various candidate legume species, including cowpea, bambara and groundnut, under the prevailing conditions, and used to parameterize the model. A number of scenarios were tested, with the objective of identifying management options with the most potential to improve legume contributions to soil fertility. The model indicated that for the legumes to be able to make any contribution of N to the system, there should be no grazing or burning of legume residues, although as cattle are an integral part of the system, this is unlikely to be a popular option with the farmers. Increasing the legume plant density to the point where it begins to affect the growth of the pearl millet will only contribute about 4 kg N ha^{-1} to the system, which may increase millet yields by 80 kg ha^{-1}, a somewhat insignificant amount. The conclusions of the study were that grain legumes alone are unlikely to be able to substantially improve soil fertility in the semiarid conditions of northern Namibia. External fertilizer inputs seem to be necessary to improve soil fertility, although the uncertain rainfall makes investment in soil fertility unattractive for farmers there.

Also in southern Africa, Prins et al. (1997) developed the KYNO-CANE model using a large database of sugarcane fertilizer trials carried out by the South African Sugar Association Experiment Station (SASEX). The model incorporated a range of factors including the influence of rainfall, soil N mineralization potential, geographic position, and the inherent soil fertility status, and could be used to derive N, P and K fertilizer recommendations based on basic economic principles. Using current (1996/97) cost:price ratios and median representative response curves, the model predicted recommendations almost identical to those of the Fertilizer Advisory Service of SASEX. However, the authors concluded that if risk scenarios were accommodated and/or a major change in the sucrose price occurred, significant differences in recommendations would be predicted.

In terms of limitations of models in this area, Hasegawa et al. (1999) report that, in the N module used by the Decision Support System for Agrotechnology Transfer (DSSAT) models, the flushes in soil inorganic N immediately after rain or irrigation when a legume cover crop was incorporated were substantially underestimated. Also the nitrification capacity factor, a variable to account for the lag between the rapid increase in ammonium concentrations and the slower nitrate release after incorporation of green manure, performed poorly. Similarly, the model failed to simulate a decrease in soil inorganic N contents after maize straw with a high C:N ratio was incorporated. Another limitation of many of the models, which is not a fault of the model itself, is the need to have accurate estimates of soil mineral N at the start of the simulation

The N module used by the DSSAT models for the simulation of soil organic matter dynamics is based on the PAPRAN model (Seligman and van Keulen, 1981) which was originally designed for high-input agricultural systems, where SOM is generally not considered of great importance for the nutrient supply of a crop. Its use, therefore, for simulating systems where SOM and residues are the only sources of nutrients for crops may not be the most appropriate, particularly as it assumes only one SOM pool. Moreover, the PAPRAN model assumes that all crop residues are incorporated into the soil, and does not distinguish a residue layer on top of the soil. However, low-input systems are often legume-based and may have a thick layer of senesced shoot parts on top of the soil. Although never incorporated, this residue may be the main source of N for the next crop. Some of these problems have been addressed by Gijsman et al. (1999) by replacing the PAPRAN module with routines from the more comprehensive SOM CENTURY model (Parton et al., 1994), although this version of the model is still being tested and is not yet widely available. CENTURY subdivides SOM into three pools, passive SOM, slow SOM and active microbial SOM, and includes two litter layers, both on top of the soil, and in the soil. Testing this modified model against a set of experiments on legume residue decomposition in Brazil described previously (Bowen *et al.*, 1992) showed good agreement between observed and simulated data (Gijsman *et al.*, 1999). The use of the CENTURY module may be limited in highly weathered soils, however (Gijsman *et al.*, 1996).

Again, it is difficult to assess whether any of these applications have had, or will have, any significant impact. The example of McDonagh and Hillyer (2000) in Namibia has the potential to focus research on soil fertility in that area on a more integrated approach involving external supplies of N fertilizer rather than on N-fixing legumes alone. The authors also suggested that the search should continue for other legume species with considerably higher rates of N fixation than cowpea, bambara or groundnut, although how successful this is likely to be is open to question as rates would have to be an order of magnitude greater to have any appreciable effect. In another example, van Keulen (1975) carried out a simulation study of growth in semiarid conditions in Israel and the Sahel, where in many years, production was shown to be limited by nutrient deficiency rather than lack of water. These findings set the stage for subsequent comprehensive research projects on primary production in both of these regions.

The ORYZA_0 example (ten Berge *et al.*, 1997b) certainly stimulated much discussion between collaborators and at workshops within the SARP project, and introduced national researchers to the use of models in their work. However, the likelihood of any change in farmer practice as a result of the work is remote and, to our knowledge, is undocumented.

Interestingly, the analysis on response farming (Keating *et al.*, 1993) indicated that only long-term analysis (25–30 years) is likely to give a clear indication of the risks associated with alternative N fertilizer application strategies. Since long-term experimental evaluation of conditional and/or alternative strategies is not possible, crop simulation models can be used to simulate alternative management strategies to evaluate the long-term risks. This approach adds value to limited experimental results in terms of evaluating constraints and opportunities for improving natural resources use and crop productivity. This has implications for the integration of crop modelling as a complementary tool into traditional agronomic research programmes, and is discussed further in Chapter 14 with reference to a specific example of the dangers of basing decisions on limited field data.

4.6 Pest and Disease Management

Because of the complexity of relationships between crops and their pests and diseases, and because populations of pests and diseases are dynamic in nature, a systems analysis approach is required to understand how pest and disease problems arise and how they may be tackled. In the case of pest and disease populations themselves, the approach bridges the gap between knowledge at the individual level and understanding at the population level. Teng *et al.* (1998) has argued that in the context of developing countries, all research on pest, disease and weed management can be viewed as ultimately contributing to the development of decision support systems for specific pests or pest complexes in a single crop or cropping systems. Examples of such decision support systems are discussed in more detail in Part 2; here we have limited our discussion to various models and their applications in the research process.

Pests and diseases can be classified into various categories depending on the type of damage caused to the crop. Boote *et al.* (1993) suggested the following classes: stand reducers, photosynthetic rate reducers, leaf senescence rate accelerators, light stealers, assimilate sappers, tissue consumers and turgor reducers. A given pest may fall into more than one of these categories, and consequently, its damage may be coupled at more than one point to a crop growth simulation model. Teng *et al.* (1998) present a framework for coupling pest and disease dynamics models to crop growth models with 21 coupling points. These were either at the levels of inputs (e.g. water, light, nutrients), rate processes (e.g. photosynthesis, transpiration/water uptake, senescence), or state variables (e.g. numbers of organs or mass of various tissues).

In terms of applications, Boote *et al.* (1993) outline an approach whereby scouting data on observed pest damage is input via generic pest coupling models in order to predict yield reductions from pests. The approach was demonstrated with the SOYGRO model to predict the effects of defoliating insects, seed-feeding insects, and root-knot nematode, with the CERES-RICE model for leaf blast and other pests, and with the PNUTGRO model for leafspot disease. They suggest that the system could be used in pest loss assessment, for projecting the effects of damage or pests on future yield, and for determining sensitivities of crops to timing of intensities of damage in specific environments. No evidence is presented, however, of use of the model or its output in any of these ways.

Rice appears to be the tropical crop on which the majority of work has been done in recent years in incorporating pest and disease effects into crop models. Pinnschmidt *et al.* (1990) coupled simple pest population models to CERES-RICE to investigate the effects of different stem-borer control strategies on yield loss. An early control with an insecticide with 80% 'knock-down' effect resulted in less yield loss than a late control. Using representative prices for rice grain and the pesticide, they found that the highest gross margins in this particular case were obtained with three pesticide applications. Calvero and Teng (1997) coupled the ORYZA1 rice crop model to the BLASTSIM.2 rice blast simulation model and a fungicide submodel (FUNGICID) to evaluate 92 spray strategies intended to manage rice blast disease. Results suggested that blast management could be obtained by limited sprays during a growing season, and that combining different fungicide types would be feasible. They also showed that a spray management scheme can be season-specific, and depends on when the infection occurs. In a subsequent paper, Teng *et al.* (1998) proposed that the use of iso-loss curves derived from pest progress curves with different onset times and maximum level reached can be used in pest management decision making to determine which kind of pest scenario needs to be controlled. In another study, Watanabe *et al.* (1997) modified the MACROS model to simulate the effect of brown rice planthoppers which are sap-suckers. They showed that variations in yield occurred with different infestation and climatic conditions, even though the initial plant biomass and planthopper infestation levels were stable.

In relation to biological pest control, van Lenteren and van Roermund (1997) note that modelling has always played a role in the process of selecting and improving the efficacy of releases of natural enemies. They give an example of how modelling approaches were able to provide insight into why control in some greenhouse crops was good (e.g. tomato) while in others (e.g. cucumber) it was not, and argue that integration of both experimentation and modelling is the way forward.

The main impact of these studies appears to be in providing insight into the complexity of the interactions between crops and their pests and diseases. Until about the mid-1980s, most damage functions used to predict crop loss in the decision making process in pest management were empirical regression equations specific to the cultivar and location of the experiment (Teng *et al.*, 1998). As such they could not be used when weather, crop cultivar, soil type or management practices changed, and were thus limited in their extrapolation value. The mechanistic approach offered by combined pest and crop models offers a more generic way of estimating these damage functions. It is always difficult to know how much of this information finds its way into practical uses, and when it does it is easy to forget or underestimate the role that modelling studies played. De Moed *et al.* (1990) give a good illustration of this for glasshouse chrysanthemums in the Netherlands. Simulation studies with a comprehensive simulation model of the epidemiology of the insect baculovirus SeNPV in populations of beet army worm (*Spodoptera exigua*) in glasshouse chrysanthemums indicated that split applications of intermediate dosages of this virus were more effective against the pest than a single high dosage. The reason was that caterpillar stages are not closely synchronized, and, therefore, the larval stages that are vulnerable to a virus spray occur over an extended period of time. The simulated positive effect of multiple sprayings led to empirical field tests of such an approach. Split applications are now standard practice. The model itself is not used for tactical decision support on site, because it requires data that are too difficult to gather routinely, but it did provide insight into the processes involved that was subsequently made use of. The SIRATAC (Macadam et al., 1990) and EPIPRE (Zadoks, 1981) pest management models are similar examples, and are discussed in more detail in Chapter 9.

A limitation of the existing models is that they generally only simulate the effects of a single pest or disease. It has been argued that multiple pest situations might give different yield losses than just summing up the effects of losses from individual pest species (Teng et al., 1998), in which case the decision criteria in pest management would be altered. However, different studies have shown conflicting results in this area, and further research is needed. Another limitation of the coupled pest–crop models as a system is the difficulties involved in actually predicting pest outbreaks. Most studies have used outbreak scenarios or used observed data as inputs into the model.

4.7 Weed Management

Weeds generally reduce crop yields through competition for limited resources such as light, water and nutrients, although in some cases, they may also release chemicals which can suppress germination or growth of the crop (allelopathy). Progress in modelling crop–weed interactions has been slow, partly due to the difficulty in designing experiments to clarify the mechanisms involved. For example, it has not been easy to separate out the relative effects of competition and allelopathy, or to determine whether above-ground or below-ground interference is the more important (Teng et al., 1998). Much work to date, therefore, has concentrated on quantifying the reduction in final yield brought about by a particular weed population without concern for the actual mechanisms involved.

Nevertheless, some progress in understanding and modelling these mechanisms has been made. Many early weed models considered only competition for light (e.g. Rimmington, 1984) by dividing the canopy into a number of horizontal mixed and unmixed layers according to the heights and canopy structure of the component species. Barbour and Bridges (1995) describe the incorporation of a weed competition module into the DSSAT PNUTGRO model based on competition for light only. Other models, such as WEED-CROP (Spitters and Aerts, 1983) and the INTERCOM model derived from it (Kropff, 1988), simulated competition for both light and water by considering the vertical distribution of leaf area and root mass. Relative competitive abilities were based on the relative leaf areas and root masses of each species. These concepts were subsequently expanded to include N (e.g. Graf and Hill, 1992; Kiniry et al., 1992). Graf and Hill

(1992) divided weeds into six groups based on differences in leaf shape, growth form, height and phenology, and calculated a potential N uptake rate for each group. Available N was then apportioned according to the proportion of root space explored by each group.

Lundkvist (1997) has reviewed and classified weed models into those used in research and those used as practical tools. Research models aim at a deeper understanding of weed–crop ecology processes and are therefore often narrowly focused on one crop and one or two weed species. Some of these models simulate and predict long-term changes in a weed population while others concentrate on specific parts of the weed life cycle. Examples of processes described by research models are: crop–weed competition, population dynamics, movement of herbicides in the soil, dose–response relations and herbicide resistance. Practical models, on the other hand, generally deal with broader systems than research models, covering a range of crops, weed species and control methods, often with low resolution. The weed control decision aids in practical models are normally based either on herbicide efficacy or cost–benefit considerations.

Examples in the literature of applications of models in either of these categories in developing countries are limited. Lindquist and Kropff (1997) used the INTERCOM competition model to evaluate the influence of early crop vigour in rice in its tolerance to barnyard grass, only considering competition for light. Results suggested that increasing early leaf area expansion and height growth rates in rice will reduce yield loss, confirming what is generally known already. Similarly, Bastiaans *et al.* (1997) used INTERCOM to show that that competition for light between rice and weeds is mainly determined by morphological characteristics, of which early relative leaf area growth rate, early relative height growth rate and maximum plant height were the most important. Caton *et al.* (1999) used the DSRICE1 model, which simulates the growth of direct seeded rice, to predict the effect of delayed rice seeding on yield reductions due to weed competition. This is important, as one of the problems associated with direct sowing of rice compared to transplanted rice is the greater predominance of weeds. VanDevender *et al.* (1997) developed a mathematical model based on Richards equations to predict rice yield reduction as a function of weed density and duration of competition. Although the authors claimed that predictions from the model should be useful in assessing the economic impact of weeds and in determining the feasibility of alternative weed control treatments, no examples of such applications were given.

An example of how crop modelling can complement a plant breeding programme is given by Dingkuhn *et al.* (1997) at the West African Rice Development Association (WARDA) in West Africa. They used the ORYZA1 rice model to explore the potential of early ground cover and high specific leaf area (SLA[2]) as factors in weed competitiveness in crosses between

[2] Specific leaf area is the area of leaf per unit weight of leaf, with units $cm^2 \ g^{-1}$, and is correlated with the leaf thickness.

cultivated rice, *Oryza sativa,* and the wild rice species *Oryza glaberrima.* The simulation results showed that high SLA in the vegetative phase and lower SLA during the reproductive phase was the best combination. Individuals with this characteristic exist amongst the progenies from the crosses, and selection trials are currently under way at WARDA.

Van der Meer *et al.* (1999) used the PARCHED-THIRST model to evaluate four weed management scenarios based on farmers' access to draught animal power and labour, taking into account ability to winter plough, time of planting and frequency of weeding during the season, for Zimbabwe. Winter ploughing was shown to reduce use of soil water by weeds so that there was more available at the start of the rainy season, but this did not appreciably affect yields. A delay in planting resulted in poor crop establishment, so that the effect of weed competition was increased.

An interesting use of a plant growth model in weed research is provided by Smith *et al.* (1993) to investigate the potential of *Smicronyx umbrinus* as a biocontrol agent of the weed *Striga hermonthica.* They developed a simulation model of the weed's population dynamics for a millet cropping system in Mali. It was found that, if *S. umbrinus* was used as the sole control agent, it would have to destroy approximately 95% of *Striga* seeds each year to reduce the emergent plant density by 50%. Such a high rate of seed destruction would be necessary because the size of the *Striga* seed bank does not limit the parasite density in most situations. Crop rotation of 1 year of millet followed by 3 years of fallow or another crop would reduce the emerged equilibrium density of *Striga* to 61% of that with no control, and destruction of an additional 22% of seeds by *S. umbrinus* would decrease this to 50%. *S. umbrinus* in conjunction with weeding would not become effective until the seed bank was limiting at very high levels of weeding.

Few examples exist of crop models being used to evaluate the effect of herbicides on crop growth in a mechanistic manner. Whisler *et al.* (1986) used the GOSSYM model to analyse the response of cotton crops to herbicide damage. To identify causes of declining cotton yields, simulations were conducted to study the effect of root inhibition and reduction in the permeability of roots to water and nutrient uptake on growth, development and lint yield of cotton under different rainfall and temperature patterns. Results showed that root inhibition actually increased yields slightly due to the redistribution of dry matter within the crop, but that decreased root permeability substantially reduced yields and delayed maturity.

The main limitation of weed models so far is that they include only a small subset of all of the factors that affect crop yield loss through weeds. For example, crop variety, planting date, row spacing and crop stand all affect a weed's ability to compete with a crop, but quantifying these effects through field experimentation alone is difficult, due to the large numbers of weed species that can occur in a field, and the diversity in growth habits of different species. Moreover, extrapolation of field results from one location to another is complicated by the fact that many weed species are photoperiod-sensitive. Competitive ability may change during a season, as plants shift from the vegetative to the reproductive phase. More sophisticated dynamic models of weed–crop competition need to be developed to help interpret experimental results and investigate possible mechanisms of crop–weed interaction.

A second limitation of most of the models described above is that they assume a homogenous horizontal distribution of crop and weed leaf area. However, a number of studies have shown that crop yield increases as distance from a weed increases (e.g. Monks and Oliver, 1988), so that such models show deviations from observed data when weed distribution in a field is not uniform, or weed densities are low. Weed distribution is likely to be clumped (e.g. Thornton et al., 1990) rather than distributed randomly. Assuming a homogenous distribution rather than confining it to an area of distribution overestimates light interception. Wiles and Wilkerson (1991) attempted to address these problems in developing the SOYWEED/ LTCOMP weed model based on the SOYGRO crop model. It was designed to simulate low weed populations near economic thresholds and considered the differential competition of a weed with crop plants at varying distances. However, it was limited by its ability to simulate only one weed species at a time.

The impact of these models so far, particularly in the developing country context, has probably been minimal, perhaps surprisingly in view of the major problems that weeds cause in many subsistence agricultural systems. They have helped to some extent to provide quantitative insight into the nature of the interactions between the crop and weed species, but seem in many cases to have only been able to confirm what is generally known already. The example of Dingkuhn et al. (1997) in identifying traits associated with competitiveness for selection in a breeding programme indicates the potential of such models to make a useful contribution in the future. Similarly, quantifying the yield loss from delayed direct sowing of rice (Caton et al., 1999) and evaluating different weed control strategies (Smith et al., 1993) can provide useful information for farmers, although to what extent this information has been disseminated to, and taken up by, farmers is not known. Evaluation of other crop management strategies on weed control such as tillage techniques (e.g. Mohler, 1993) and crop rotations (e.g. Jordan et al., 1995) also need to be explored in a tropical agriculture context.

As far as future development of models for weed research is concerned, Doyle (1991), in reviewing the use of models in weed research, considered that there has been a tendency to concentrate on certain aspects, such as defining threshold levels and weed–crop competition, to the exclusion of other critical areas. In particular, he cites weed dispersion, variation in recruitment and mortality, spatial heterogeneity in weed populations and the existence of multi-species assemblages as areas that require more attention.

4.8 Harvesting

There are not many examples of how crop models have been used to determine optimum harvest date, presumably because for most crops the date of harvest is determined by the physiological maturity of the crop and prevailing weather conditions, both of which a farmer has little control over. Nevertheless, for some crops, decisions need to be made when to harvest, and some models have been developed to determine the best time to do so.

For example, if wheat is harvested too early, the moisture content of the grain will be too high and the farmer will need to dry it with expensive equipment. On the other hand, if harvesting is too late, summer storms can delay harvesting still further and yield and quality losses occur. Abawi (1993) developed a model to optimize crop harvesting date for wheat in Australia taking these factors into account, although it was not stated if the model was taken up and used in a practical way.

Also in Australia, limited mill capacity often means that it is not possible to harvest all sugarcane when maximum yields are attained. A model was therefore developed to maximize sugar yield and net revenue in relation to harvest date and crop age (Higgins et al., 1998). Results showed that there was scope for optimizing harvest date to improve profitability given current harvest season lengths and land area. Similarly, Martinez Garza and Martinez Damian (1996) developed a simulation model to forecast harvesting date for sugarcane based on temperature and precipitation data which gave good predictions of appropriate harvesting dates over a 38-year period for Tamaulipas in Mexico.

For pineapple, Malezieux et al. (1994) describe improvements to a model for predicting pineapple harvest date in different environments, to aid scheduling of labour and fresh fruit marketing efforts. The original version of the model, developed for pineapple in the Smooth Cayenne group could not satisfactorily predict fruit harvest date in the range of environments in which pineapple is grown in Hawaii. The improved model was able to predict harvest dates for pineapple growing in Australia, Côte d'Ivoire, Hawaii and Thailand, within a margin of error that would be acceptable to most pineapple growers, although it is not stated whether growers actually use it or not.

Lu and colleagues (Lu et al., 1992; Lu and Siebenmorgen, 1994) took the approach of trying to model moisture content of rice grain throughout the harvest season. Attempts were then made to relate final yield, percentage head rice and percentage milled rice to the moisture content. The model, using both hourly and daily field meteorological data, was able to predict moisture content with reasonable accuracy. There is, however, no evidence of its use in a practical situation.

Cropping and Farming Systems

5

Robin Matthews

Institute of Water and Environment, Cranfield University, Silsoe, Bedfordshire MK45 4DT, UK

In the previous chapter, we considered examples of how crop–soil simulation models had been used to further understanding of the management of a single crop. There is often a tendency amongst crop modellers not to think further than the crop of their interest, and to forget that, not only is that crop probably only one of several that are being grown by a farmer, but also that crops as a whole may be only one activity amongst several others in an overall farming system. Many other enterprises, such as livestock, aquaculture and vegetable production, may also be important components of the way in which farming families sustain their livelihoods, and in many cases, cropping may not even be the main source of income for a household. In this chapter, we consider how crops fit, first of all into cropping systems, and secondly, into overall farming systems, and describe attempts that have been made to use crop simulation models in association with models of other farm components.

5.1 New Crops and Cropping Systems

Agriculture is constantly changing, both through innovations from within and also through the influences of outside forces. Researchers and planners are often interested in the feasibility of growing new crops in a region, while farmers may sometimes be forced to adapt to comparatively rapid changes in their biophysical and socio-economic environments. In both cases, where there is a degree of uncertainty, modelling offers a way of exploring long-term impacts quickly and effectively in a way that experimentation cannot do.

For example, Fukai and Hammer (1987) developed and used a model to estimate productivity of cassava at different locations in northern Australia. Cassava was being considered as a promising crop in that region for the production of ethanol which could be used to supplement petrol, and suitability of the crop under different environments needed to be evaluated. The model was developed to extrapolate results from a series of field experiments throughout the region. The results identified that the two factors limiting cassava were low temperatures in the winter in the southern area, and low rainfall in general. The information provided guidelines regarding the establishment of the cassava industry in the region. A similar study was carried out for sorghum (Hammer and Muchow, 1991), in which the production risk due to spatial and temporal variability was quantified at different locations. It was suggested that the results would be of use for planning and policy making in the sorghum industry, although it is not known if this has occurred since then.

In the Philippines, Penning de Vries (1990) refers to work using a model to evaluate suitability of soybean, where it is a new crop. Yields for four situations (rain-fed upland, irrigated upland, rain-fed lowland, irrigated lowland) were simulated over 20 years, and costs and benefits were analysed to give potential net profit. Results showed that even when rice remains the first crop planted, there is a window of 2 months to grow soybean profitably in deep soils. On shallow soils, however, this window lasts only a few weeks. No evidence is given that this information was taken up.

An interesting use of crop models to evaluate changes in cropping systems is given by Thornton *et al.* (1995a). In Colombia, in an attempt to boost the price of coffee, the government made a one-off cash payment to induce coffee farmers to move to annual crops such as maize. However, there was some concern that as farmers' planning horizons shifted away from the long term more to the short term, soil erosion might increase. Thornton *et al.* (1995a) used a whole-farm model incorporating several individual crop models to evaluate various crop management options for maize-based systems after conversion from coffee. The results showed that a maize–bean–tomato–cassava rotation was the best rotation in terms of highest mean income and lowest levels of risk. However, although short-season annual crops were more attractive than traditional perennial crops in the short to medium term for individual farmers, the consequences were unacceptable social costs associated with degradation of soil and water supplies. The authors make the point that it is essential to carry out the analysis of the same problem at different scales, as this is often the root of conflicts. However, there is no evidence that the findings of the study were taken up by the relevant policy makers in this case.

In Argentina, Savin *et al.* (1995) used the CERES-WHEAT model to investigate cropping strategies for rain-fed wheat involving the combination of two cultivars of different maturity (early and intermediate) and two sowing dates (early and late), in two different locations in the Pampas. Simulation experiments using 24 years of daily weather records showed that the greatest stability of grain yield was obtained if the early-maturing cultivar was sown late. However, if N was not limiting, maximum yields could be achieved by early sowing of the intermediate maturing cultivar, although this strategy also produced the lowest yield in the worst years. When fallows were not weed-treated, the yield of the early-maturing cultivar was affected more than the intermediate one, particularly in years of overall low yield.

In another example, Keatinge *et al.* (1999) used a simple crop phenology model and calculated temperatures to examine the suitability of six legume cover species for use in farming systems of the mid-hills region of Nepal. The criteria was that they must reach maturity prior to the sowing period for the principal summer cereal crops. Results showed that *Vicia faba*, *Vicia villosa* ssp. *dasycarpa* and *Lupinus mutabilis* would be suitable as autumn-sown crops across most of the mid-hills if early sowing is possible. *Vicia sativa* and *Trifolium resupinatum*, on the other hand, are only likely to mature early enough at lower elevations. Similar exercises were conducted for hillside regions in Bolivia (Wheeler *et al.*, 1999) in which potential cover crops, not grown locally, were recommended for further trials, and in Uganda (Keatinge *et al.*, 1999), the results of which were taken up by CARE International in designing field trials. It was concluded that the models were useful tools to pre-screen a wide range of legume genotypes to eliminate unsuitable germplasm from further field testing, and had potential for scaling up field tests to produce suitable recommendation domains.

The work already discussed in Chapter 4 (Section 4.5) on nutrient management is another example of how models can be used for an initial evaluation of new crops in a region. McDonagh and Hillyer (2000) describe how a model was developed to evaluate whether legumes intercropped with pearl millet could help improve soil N status in northern Namibia. A number of legume species, including cowpea, bambara and groundnut, were tested, with field measurements of their rates of N fixation being incorporated into the model. Results of the study indicated that these legumes by themselves are unlikely to be able to improve soil fertility substantially in northern Namibia, and that external inputs of fertilizer seem to be necessary.

Robertson *et al.* (2000) describe an innovative participatory approach involving researchers, farmers, advisers and grain traders to explore yield prospects for a spring-sown mung bean crop in October/early November after a winter fallow. Mung bean is traditionally seen as a low-yielding high-risk crop by farmers in northern Australia, primarily as it is usually sown in December/January after a winter cereal, when soil water is low. Previously, spring sowing was not recommended due to weather damage in the winter leading to low prices in a market demanding high quality. However, with new varieties available, and the opening of a market for intermediate quality grain, these recommendations needed to be rethought. The key elements of the approach described were to use simulation studies to identify possible options, to test new practices with innovative farmers, and to monitor the management and performance of commercial crops to compare yields with benchmarks estimated with the model. After two years of on-farm testing, spring-sown mung bean was shown to have a potential for high returns.

A limitation of the approach of using crop models for initial evaluation of new crops and cropping systems is often the lack of weather and soils data for the particular region for which the new crop or cropping system is being evaluated, particularly in developing countries. However, various techniques now exist for interpolating scarce data (e.g. Chapman and Bareto, 1996), and also the collection and collation of environmental data in developing countries is improving all the time, so this limitation should decrease in significance as time progresses.

Some of these applications have had impact. The example given by Robertson *et al.* (2000) for mung bean in Australia, indicating that due to changed external circumstances (i.e. new varieties and markets) a new cropping system could be contemplated, clearly had some impact – all but one participating farmer out of 22 was keen to grow the crop again. There has also been a rapid rate of adoption of the overall research approach since it was first publicized in 1996. Similarly, the results of the evaluation of leguminous cover crops in Uganda have been used by CARE International in their field trials (Keatinge *et al.*, 1999). Although not yet documented, the Namibian study has the potential impact of influencing soil fertility research in the semiarid areas of that country. By showing that N fixation by legumes is unlikely to be able to supply adequate N to sustain soil fertility under arable crops, research can be focused on more productive ways of maintaining or improving soil fertility. The Australian studies on cassava and sorghum produced useful information for planners in the respective industries, but it is not known to what extent this has been made use of.

5.2 Evaluating Sustainability

There is little consensus on what sustainability is, but most definitions contain the concept of time, usually the long term, and some measure of the performance of biological, environmental or socio-economic systems. Sustainable agricultural production systems should meet the requirements of the farm household in terms of food, income and leisure without endangering the productive capacity of the natural resource base. To assess the degree of sustainability of a particular system, there is a need to understand quantitatively the processes determining production and how these are influenced by soil characteristics, environmental conditions and management practices. Long-term experimentation is one way to gather the required information; such experiments do exist in the UK (e.g. Rothamsted, ~140 years), the USA and elsewhere, and can give valuable insights into soil fertility issues and the sustainability of crop yields. However, these experiments are the exception rather than the rule, and have their limitations – they are laborious and time consuming, and generally take too long to give results within the timeframe to make decisions. Moreover, variability in environmental conditions makes it difficult to use results specific to one time and place for extrapolation to other environments (van Keulen, 1995).

Crop–soil models offer a cheaper and quicker complementary approach, and can easily evaluate a number of alternative strategies in terms of their sustainability. Models are able to indicate future trends and prescribe appropriate action to minimize harmful effects, such as the use of suitable crops, adapted varieties and changes in management and cultural practices. The following are some examples of how such models have been used to assess the sustainability of various cropping systems.

Schipper *et al.* (1995) used a linear programming approach to evaluate the effect of a number of policy interventions in Costa Rica on land use and the sustainability of soil nutrients (N, P, K) and a biocide index, assuming that the goal of each household was the maximization of farm income. The objective functions within the analysis were determined in part by crop simulation models. The analysis took into account five different farm types based on their land:labour ratios, and proportions of three different soil types. Policy interventions investigated included changes in output, input and factor prices, capital availability, and biocide regulatory measures. Results showed that for a number of these policy instruments, there were trade-offs between sustainability and income objectives.

Hsin-i *et al.* (1996) used a modelling approach to address increasing crop and forage production in a semiarid agroecosystem in north-western China. Using a proposed rotational scheme for winter wheat and lucerne production, portions of arable land were allocated for raising animal forage in order to improve soil productivity. Animal manures were returned to the soil to stabilize the structure further and to maintain fertility. Projected increases in population were also taken into account. Simulation results over 40 years indicated that in all cases, soil quality was improved by the use of crop rotation and animal manure. Long-term application of N fertilizer as an alternate management practice was evaluated using the model, but detrimental effects on soil quality were predicted, mainly due to the stimulation of grain yields which removed trace elements from the soil. The authors express the hope that the message that the beneficial effects of using crop rotations and manures would be disseminated over the region, and that the government would provide economic incentives for its adoption, although the mechanism for the uptake of the information is not described.

To answer the question of whether to take crop residues from, or to leave them in fields in Niger, Bruentrup *et al.* (1997) compared the long-term economics of using a crop residue mulch to those for the complete removal of crop residue. Results showed that not only was short-term profitability low from using the crop residues as a mulch, but also with mulching alone, future soil degradation could not be prevented. A regular fallow period for long-term fertility maintenance was necessary, but if the fallow was to be shortened, it was important that mulching was practised.

In Colombia, Hansen et al. (1997) used a farm model linked to process-level crop models to characterize determinants of sustainability of a hillside farm in the upper Cauca River. Sustainability was expressed as the probability of farm continuation. Results identified the cropping system, the area under cultivation, consumption requirements and crop prices as important determinants of sustainability. The impact of price variability and spatial diversification on farm risk were also shown to be important. The approach was used to evaluate different ways that had been suggested of enhancing the sustainability of the farm. Results showed that farm sustainability could be enhanced by intensifying and diversifying production of annual crops and by including tomato and other high-value vegetables in the crop rotation. The authors also identified some limitations of the crop models used, which included over-prediction of yields, underestimation of year-to-year variability and inability to simulate several yield-reducing stresses important in that environment.

Timsina et al. (1997) used the CERES-WHEAT and CERES-RICE models run in series to simulate long-term sequences of yields in the rice–wheat system in northern India. The results from the simulations matched experimental results showing that rice yields have declined and wheat yields have increased over the period. They then used the models to identify the causes for low unstable yields, and to quantify nutrient depletion rates in the system.

Hedgerow intercropping has received considerable attention in recent years as a potential means to help maintain soil fertility and reduce soil erosion. Nelson et al. (1998b) used the Agricultural Production Systems Simulator (APSIM) model to evaluate the sustainability of maize crop management practices in the Philippines. With no hedgerows present, continuous maize cultivation was predicted to be unsustainable in the longer term, although the inclusion of a fallow period was predicted to slow the productivity decline by spreading the effect of erosion over a larger cropping area. Hedgerow intercropping was predicted to reduce erosion by maintaining soil surface cover during periods of intense rainfall, contributing to sustainable production of maize in the long term. However, insecure land tenure limits the planning horizons of upland farmers, and high establishment costs reduce the short-term economic viability of hedgerow intercropping relative to open-field farming (Nelson et al., 1998a). In the long term, high discount rates and share tenancy arrangements in which landlords do not contribute to establishment costs make hedgerow intercropping even less attractive.

The same APSIM models were used by Probert *et al.* (1998) to simulate the performance of a hypothetical chickpea–wheat rotation on clay soils in Queensland. The results indicated that soil organic matter and N steadily declined over 25 years in a continuous wheat cropping system without the addition of N fertilizer, whereas the inclusion of chickpea in the rotation considerably reduced the decline in soil fertility. The legume effect was evident in the amount of nitrate available at the start of each wheat season. Similar results were obtained by Bowen *et al.* (1998) using the Decision Support System for Agrotechnology Transfer (DSSAT) models to investigate the long-term sustainability of various cropping systems in Brazil (Fig. 5.1). A continuous maize–fallow system with no inputs of fertilizer caused maize yields to decline gradually over 50 years, whereas a green-manure–maize–fallow system was able to maintain yields over the same period. However, this source of N from the green-manure was not adequate for maximum yields.

Probert *et al.* (1995) used the CENTURY and APSIM models to examine the effects of tillage, stubble management and N fertilizer on the productivity of a winter–cereal–summer–fallow cropping system in Australia. Both models predicted that for this continuous cereal cropping system, there would be a decline in soil organic matter and a reduction through time in the capacity of the soil to mineralize and accumulate nitrate during the fallows. Yield predictions with APSIM were sensitive to carryover errors in the water balance from one season to the next, so that in some seasons large errors

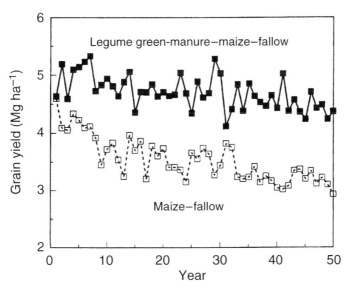

Fig. 5.1. Expected grain yields obtained from simulating a 50-year short-duration legume green-manure–maize–fallow rotation compared to a maize–fallow rotation with no N inputs for a site in central Brazil (from Bowen *et al.*, 1998).

occurred in the predicted relative yields. Both models reproduced the observations well enough to indicate their suitability for studying the behaviour of cropping systems where the focus is on depletion of soil fertility.

Menz and Grist (1998) describe the use of the SCUAF (Soil Changes Under AgroForestry) model to evaluate the influence of burning and changing the length of the fallow period in Indonesian shifting cultivation systems. Output from the model was input into a simple cost–benefit spreadsheet model to calculate the economic viability of the different options in terms of returns from rice. Results indicated that, as the length of the fallow period declined from 20 years down to continuous cropping (i.e. no fallow), rice yields and the levels of C and N in the soil would all decline, and that the amount of soil lost due to erosion would increase due to reduced vegetative cover. However, the economic analysis showed that maximum revenue (over a 63-year period) came from a 1-year fallow system – even though yields were lower, what was obtained was still able to more than compensate for the loss of productive land with longer fallow periods. They concluded that even though more intensive forms of cultivation carried a future yield penalty, it was not sufficient to overcome the more immediate economic gains to be made from intensive cropping. Burning to clear the land for cropping resulted in faster decline of soil carbon levels, not only due to the reduction in C being added from the vegetation, but also because of enhanced soil erosion from decreased ground cover. Burning had approximately the same effect on soil fertility as a 1-year reduction in fallow length, but was the most profitable method of clearing land for cropping due to the low labour and cash requirements compared with other options such as clearing manually or using herbicide. This matches current farmer practice. The authors make the point, however, that off-site costs of burning, such as smoke haze, CO_2 release to the atmosphere and silting up of water courses, were not factored into the costs. Like many others, they conclude that such studies are of value for policy makers, but do not state if the information has been used in this way.

The main limitation in using crop simulation models for analysis of long-term trends is that they have not yet been thoroughly validated, certainly not over a relevant time-span to judge their long-term behaviour. This is partly due to the shortage of good quality long-term data, particularly as many variables needed to validate the models thoroughly have not been measured in the data sets that do exist. However, there have been recent efforts to compare a number of soil organic matter (SOM) models against data from long-term field experiments (see Smith *et al.*, 1997b).

A second limitation is our partial knowledge of many of the underlying processes. Most crop models were originally designed to describe crop growth and soil processes over one season, and the relatively simple relationships generally employed are usually adequate for this time period. However, we just do not know whether all of these relationships are adequate to describe soil changes over much longer time periods. The

PAPRAN model used as a basis for the soil N transformation routines in all of the DSSAT models, for example, uses only one pool to represent the soil humus fraction which is unlikely to give good results over longer periods. The recent addition of the well-tested CENTURY SOM model routines (e.g. Kelly *et al.*, 1997) into the DSSAT crop models (Gijsman *et al.*, 1999) may help improve confidence in long-term predictions by these models. Error propagation within the models may be another potential problem – a small error may be relatively unimportant over a single season, but over several seasons it could accumulate and result in substantial error at the end of the run. So far, little work has been done to investigate the magnitude of such errors.

Other limitations depend on the particular model, but for the DSSAT crop models they include the need to incorporate soil erosion and tillage, the effect of certain fertilizers on soil acidity and the effect this has on crop growth, the effect of salinity on crop growth, and the retarded leaching of nitrate in the subsoil of some soils. These models also do not currently distinguish between plant residues left on the surface and those incorporated in the soil. Some of these limitations are now being addressed – for example, a salinity module for use with the CERES models is now available (Castrignano *et al.*, 1998).

Nevertheless, it is highly likely that modelling will become an increasingly important tool for the study of sustainability and environmental problems simply because there are no other reasonable approaches to quantify the complex processes involved.

5.3 Farm Household Models

Modelling at the farming household level can be traced back to the 1950s (Jones *et al.*, 1997), and led to the development of the new household economics (NHE) theory (Becker, 1965). This approach assumed that households act as unified units of production and consumption which aim to maximize utility subject to their production function, income and total time constraint. The first 20 years saw the research emphasis on linear programming techniques, during which economists analysed farm growth, response to policies, cost minimization, minimum resource requirements, etc. Throughout, there was a general assumption that farmer behaviour was governed by profit maximization. In the 1970s, there was a shift in emphasis towards the development and use of macroeconomic models for policy analysis. However, farm-scale modelling continued, with the emphasis more on the development of econometric household models that could apply to developing agriculture (e.g. Yotopoulos *et al.*, 1976; Lau *et al.*, 1978; Barnum and Squire, 1979). Some studies were conducted to determine how to aggregate farm models to regional and national results. In the 1980s, there was a growing realization that the macroeconomic models did not allow an adequate understanding of the likely effects of policies and market conditions on individual farmers, and there was a shift in emphasis back towards farm-scale analysis. At the same time, there was an appreciation that households are not homogenous units, and that inequalities amongst household members do occur. This has led to the development of a class of models called bargaining models, which recognize that household members may have different preferences (McGregor *et al.*, 2001). An example of the use of this type of model is that of Jones (1983) who analysed the mobilization of women's labour for cash-crop production in the Cameroons.

The main purpose of most of the recent studies has been to gain improved understanding of farming systems regarding response to policies, resource requirements, farm growth potential and, most recently, environmental impacts of farming practices. For example, Gundry (1994) developed an integrated model of the food system in a region in Zimbabwe to carry out comparative analysis of the factors affecting nutritional status of households in the area, to help identify households most at risk. Such a framework could be used to assess proposed policy interventions aimed at promoting food security, and to assist in targeting food aid distribution, although it is not stated whether it was actually used in this way. In another example, Lee *et al.* (1995) used a farm model to help understand the behaviour of subsistence farmers in Western Samoa after government efforts to stimulate production of exportable crops had only limited success.

The behaviour of crops in most of these models was usually based on either average yields of crops derived from historical data, or from empirical production functions relating yields to resources such as water or N. In some cases, these production functions were derived from several simulations by a crop model. However, this approach was restrictive as: (i) they required data of the particular management practice under the appropriate weather and soil conditions to be collected beforehand (inadequate if being used to investigate new locations or conditions or new management strategies); and (ii) decision makers often need to know more about a system than just yield (e.g. leaching, erosion, run-off, etc.). Moreover, these modelling approaches presuppose that the household has a 'goal' – generally assumed to be maximization of profit (Edwards-Jones *et al.*, 1998), an assumption that has been shown to be inadequate in trying to model the adoption of new technologies, such as selection of rice varieties, by households (e.g. Herath *et al.*, 1982). Day-to-day decision making by householders is not usually considered. Such models are structured to represent an equilibrium when production has stabilized – climate conditions, for example, are assumed to be average every year. They are not able, therefore, to simulate processes leading from one equilibrium to another particularly well (Pannell, 1996). Such models have also not generally considered the wider costs to society, such as pollution, or a decline in the natural resource base.

In an attempt to address some of these problems, Hansen (1995) and Hansen and Jones (1996) coupled a number of crop models to a decision model to simulate household consumption and interactions between resources. The model did not, however, attempt to simulate the decision-making behaviour of the farmer. The operation of each of the crop models was based on a management plan defined previously by the user, which controlled the sequence and management of crops for each field on the farm. The model was used to assess various farm enterprise options for their 'survivability' over 6–10 years in the face of price risk and natural resource degradation in the form of soil erosion (Hansen *et al.*, 1997).

Edwards-Jones *et al.* (1998) took this work further and demonstrated the feasibility of integrating socio-economic and biological models by linking CERES-MAIZE and BEANGRO to family decision-making and demographic models to represent a subsistence farming system. However, the work highlighted a number of issues that needed to be addressed using this approach:

1. The crop model structure did not allow tactical decision making during the season, i.e. all decisions had to be made either before or after each simulation.
2. What is the appropriate scale for modelling farm households, i.e. do individuals need to be modelled? One option suggested is a frame-based approach – each individual is represented by a frame for specific physical, social and psychological characteristics. The frames are controlled by rules and interact to reach a decision. However, this approach may require too much data.
3. Are the socio-economic variables currently being measured in surveys the most appropriate? There may be less tangible variables that are more important in determining behaviour.
4. Validation of the model is difficult.

In a recent attempt to make progress in this area, Shepherd and Soule (1998) developed a farm simulation model (Fig. 5.2) to assess the long-term impact of existing soil management strategies on the productivity, profitability and sustainability of farms in western Kenya. The model, written using the STELLA modelling package, ran in time-steps of 1 year, and linked soil management practices, nutrient availability, crop and livestock productivity, and farm economics. Crop types included weeds, fodder, grass, shrubs and two types of grain crops. Growth of the plants was determined by the N and P availability. A wide range of soil management options was simulated, including crop residue and manure management, soil erosion control measures, biomass transfer, improved fallow, green manuring, crop rotations, and N and P fertilizer application.

The model was applied to a case study of the sustainability of existing farming systems in the Vihiga district in western Kenya. Three household types were simulated to represent the range of resource endowments of

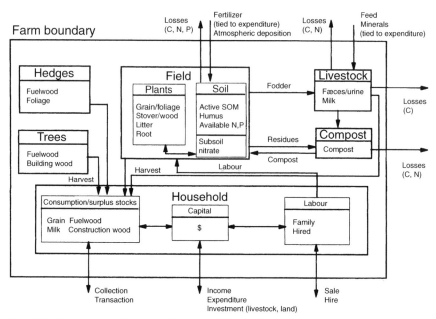

Fig. 5.2. Conceptual diagram showing the principal compartments and material flows that are represented in the farm economic–ecological simulation model of Shepherd and Soule (1998). The flows are modelled in terms of C, N, P and monetary value. Not all model flows are shown.

households in the study area. Data to initialize the model for the different household types was collected through participatory research in the area, which indicated large differences in farm size, quantity and quality of live-stock, soil and plant management, food consumption patterns and sources of income. The model was used to assess the sustainability of the existing systems for the three farm types as a basis for recommending improved practices for each. It was shown that the low and medium resource endow-ment farms had declining SOM, negative C, N and P budgets (Fig. 5.3), and low productivity and profitability. The high resource endowment farms, on the other hand, had increasing SOM, low soil nutrient losses, and were productive and profitable. This disaggregation into household types according to resource endowments highlights the dangers of relying on nutrient balances of only one 'average' farm type – most previous studies in Africa have generally shown negative nutrient balances using this approach. Nevertheless, in this particular case, the low and medium resource households represented around 90% of the total in the study area, suggesting that overall a negative nutrient budget was likely. There is also the question of where the increasing nutrients of the high resource house-holds was coming from – their greater purchasing power may have meant that there was nutrient flow to the richer households from the poorer ones.

Shepherd and Soule (1998) concluded that the ability of the high resource households to manage their farms profitably and sustainably indicates that it is possible, but that capital is required. Strategies they suggested to improve livelihoods included: (i) an increase in the value of farm output; (ii) an increase in high-quality nutrient inputs at low cash and labour costs to the farmer; and (iii) an increase in off-farm income.

Limitations of the model, discussed by the authors, included the fact that the farm was treated as a single field, although several crops could be grown in it. There was, therefore, only a single value for the C, N and P stocks, although in practice, farmers have many fields being managed in different ways. The time step was 1 year, which did not take into account seasonality in rainfall, nutrient availability, fodder quality, prices and labour demand. The crop response to nutrients was based on empirical relationships, and production could only be altered by physical changes in the soil, plant and livestock management entered by the user, and was not responsive to changes in labour inputs. Soil erosion was not considered. Nevertheless, the model represented a useful start to the dynamic modelling of whole farms.

As part of a project looking at integrated agriculture aquaculture farming systems in the Philippines, Schaber (1997) developed a whole-farm model called FARMSIM to quantify flows of nutrients between the different farming enterprises (Fig. 5.4). Again using the STELLA modelling package, he combined the ORYZA_D rice simulation model with a fish-pond model (for Nile Tilapia *Oreochromis niloticus* L.), and models of pigs, chickens and buffaloes. After the rice harvest, the straw was assumed to be composted, the bran to be fed to the pigs and to the fish, and broken rice to be used as chicken food. If this supply of food was less than the demand, then more had to be purchased externally. Weeds from the field bunds were fed to the buffaloes. Manure from the pigs and buffaloes was fed to the fish, although buffalo manure could also be applied to the rice fields. The model was then used to evaluate three different scenarios to evaluate

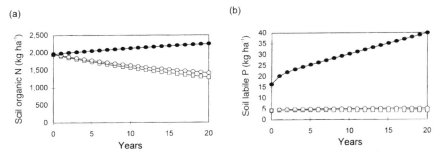

Fig. 5.3. Simulated time trends in (a) soil organic N and (b) labile phosphorus for low (open squares), medium (open circles) and high (filled circles) resource endowment categories of farms in Vihaga district, western Kenya (from Shepherd and Soule, 1998).

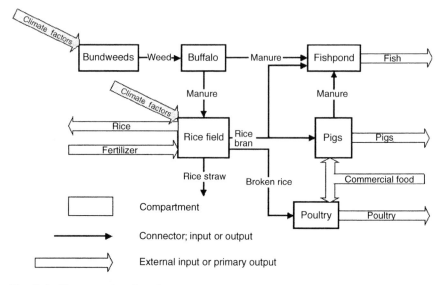

Fig. 5.4. Diagram showing the compartments and resource flows of the FARM-SIM model (from Schaber, 1997).

the efficiency of N use of the farm as a whole, defined as the output N as a ratio of the input N. In the first scenario, a conventional farming system with a monocultural rice field, two buffaloes and 15 chickens was assumed. High levels (200 kg N ha^{-1}) of commercial fertilizer were applied to the rice field, and the output of rice grain was high. In the second scenario, diversification increased as pigs were introduced. The third scenario represented a fully integrated, diversified farming system, including a fishpond, with all reusable products from each farm enterprise being used as inputs to other enterprises where appropriate. The predicted efficiencies of N use in each of the three scenarios were 13.3, 18.7 and 21.6%, respectively. It was concluded that the greater the number of enterprises there were on the farm, the greater the efficiency of N use of the farm as a whole.

As such household models are still in their infancy, it remains too early to say whether there has been any impact.

Regional and National Planning

6

Robin Matthews

*Institute of Water and Environment, Cranfield University,
Silsoe, Bedfordshire MK45 4DT, UK*

Regional and national planning involves the analysis of information from a wide area encompassing many different livelihoods and production systems, and the making of decisions to meet specified goals for the area. Questions of what is the most optimal use of land to meet these goals are therefore of prime importance to policy makers. They are also usually interested in knowing what changes their decisions will bring to the area – for example, what would be the effect of making cheap credit available for agricultural inputs on particular patterns of land use? The function of land-use planning is 'to guide decisions on land use in such a way that the resources of the environment are put to the most beneficial use for man, while at the same time conserving those resources for the future' (FAO, 1976). Basic information on soils, topography, climate, vegetation, as well as socio-economic variables such as relation to markets, skill of land-users, level of social and economic development, etc., all need to be taken into account.

The area in question can therefore be regarded as a system with inputs, products and a mix of technologies to convert the inputs into products in such a way as to meet a set of specified goals. In principle, it may be possible to produce a wide range of products, but in practice it will not be economically efficient to produce all of these products in the biophysical and socio-economic environment of a given region. The problem is in trying to define the 'best' mix of technologies; invariably, there will be conflicting views between the various actors within a region on what these are. To arrive at a well-founded choice and to be fair to all actors, van Keulen (1993) proposed three steps. The first step is to identify the range of technical possibilities in a region, as dictated by the natural resource

base and the available production techniques. The second step is to identify all possible development objectives (including non-tangible objects,
such as peoples' aspirations), and the third step is to analyse the viability
of the technical possibilities in terms of meeting these objectives within
the context of the prevailing socio-economic environment.

The central problem is how to pursue several, and often sometimes conflicting, goals which cannot be expressed in the same unit of measurements. For example, regional planners may be interested in identifying
production measures that will maximize profit for the farm producers and
minimize environmental damage to the region. All these goals are conflicting as maximizing revenue may mean high risk, mechanization (less
labour) or high chemical inputs (environmental pollution). At the national
and global levels, food security and economic profitability are major goals
in international agriculture, but ecological balance is also causing increasing concern. Progress has been slow in applying systems thinking to agricultural policy, but methodologies are currently in the process of being
developed which may prove to be useful tools to aid decision makers at
these levels. Breman (1990), for example, used a systems analysis approach
to determine the degree of exploitation of natural resources and to evaluate possible development options for southern Mali, where overpopulation
is becoming an increasing problem. An analysis of the N balance showed
that the system is not in equilibrium – outputs, which include harvests,
runoff, leaching, erosion, consumption losses and fire, are considerably
more than inputs from rainfall, legumes, manure and fertilizers.

6.1 Linear Programming Approaches

Increasing use is being made of Interactive Multiple Goal Linear
Programming (IMGLP) techniques, which allow a quantitative exploration
of the various options available in order to meet a number of goals. The
methodology of this approach is described more fully by van Keulen (1993),
but involves the use of an 'activity' matrix containing all existing and con-
ceivable production techniques for a region, including those that may still
be in the R&D pipeline. Data on production activities already practised in
a region can be obtained from statistics, but for production activities
not yet practised, crop and livestock model predictions are the most prac-
ticable source. Reiterative techniques then 'search' for the best combination of production techniques to meet the specified goals. The applicability
of the approach was demonstrated by van Keulen and Veeneklaas (1993)
to evaluate options for agricultural development in Mali. They found
that it was a good means of communication between scientists, planners and
policy makers at workshops at both the regional and the national levels.

In a major study funded by the Netherlands Scientific Council for
Government Policy, van Latesteijn (1993) used the WOFOST crop simula-

tion model and a geographic information system (GIS) containing soil characteristics, climatic conditions and crop properties for Europe, to calculate regional yield potentials for indicator crops. A linear programming model containing several policy-derived objective functions was then used to calculate the optimum regional allocation of land-use in the European Union. Four scenarios were evaluated based on issues in agricultural policy: (i) free market and free trade; (ii) regional development; (iii) nature and landscape; and (iv) environmental protection. The results of the study were used as a basis for formulating strategic choices concerning the future of agricultural areas in the region. Similar studies using crop models for agroecological characterization (described in more detail in Chapter 10) have been used by the International Potato Centre (CIP) to target global potato research to those areas where improvement in production seems most promising (van Keulen and Stol, 1995), and to estimate global food production in the year 2040 (Penning de Vries *et al.*, 1995).

In India, Selvarajan *et al.* (1997) used the ORYZA1 and WTGROWS models for rice and wheat respectively to provide the technical coefficients for a linear programming model to analyse trade-offs at the district level between expected income, income risk and water use, for a rice–wheat cropping system. Technical coefficients are inputs and outputs of production systems and include such things as yield, costs, labour use and sustainability indicators. In many cases, the objectives of farmers conflict with those of society as a whole – in this case, society's objective was to minimize water use, whereas farmers objectives were to seek satisfactory levels of average income while minimizing the standard deviation of income. At the farmer level, the conflict between income and income risk was controlled through water use options. In this way, it was shown that water use and the area under rice–wheat cultivation could be reduced without a reduction in income if other inputs (e.g. N) were increased. Pursuing maximum income with minimum income risk levels by using groundwater caused inefficiency in water use and a steady fall in the water table, thereby threatening the long-term sustainability of the system. Currently, the prices of water for irrigation and energy for pumping groundwater are heavily subsidized by the government, and do not reflect their opportunity or scarcity value, and consequently, use of water is often excessive to compensate for variations in other inputs. They concluded that rational water and energy pricing policies were needed to promote efficient water use.

In another Dutch-funded project (REPOSA, the Research Programme on Sustainability in Agriculture, described in more detail in Chapter 12) in the Northern Atlantic Zone of Costa Rica, the methodology used by van Latesteijn (1993) was further developed to quantify trade-offs between socio-economic and biophysical sustainability objectives in regional land-use studies (Jansen, 2001). The methodology that was developed – called SOLUS (Sustainable Options for Land Use) – included a linear programming (LP) model, technical coefficient generators for livestock (PASTOR) and cropping (LUCTOR) activities, and a GIS. LUCTOR (Jansen and Schipper, 1995) used a combination of a crop simulation model (MACROS) and expert judgement to define crop options according to specific crops and associated management practices. Crops included banana, black bean, cassava, maize, melina trees, palm heart, pineapple and plantain. The technical coefficients were generated using the 'target-oriented' approach, in which target production levels were predefined, and the technical coefficient generator calculated the inputs required to achieve this target. For each crop, 72 alternative management levels were defined by combining nine levels of fertilizer inputs, four types of pesticide inputs (insecticides, fungicides, nematicides and herbicides), and two levels of mechanization. Five different types of pasture with a range of different rates of stocking and fertilizer applications were also modelled, giving 420 combinations in total. Soil nutrient balances were calculated using an adapted version of the NUTMON model (de Jager et al., 1998). The LP model maximized regional economic surplus subject to a number of resource and sustainability constraints. Economic surplus and labour employment were taken as economic sustainability indicators, and soil N–P–K balances, pesticide use, nutrient losses and trace gas emissions as biophysical sustainability indicators. In the particular case study reported by Bouman et al. (1998), it was concluded that sustainable production systems could be introduced, and environmental effects reduced, if each was done separately, but if both were introduced together, then economic surplus and agricultural employment would be reduced. They found that the agricultural area could be decreased and the forested area increased without affecting the regional economic surplus.

In a subsequent analysis in the same region, individual farms were classified into five types based on resource endowments, which included farm size, the availability of land units and the land–labour ratio (Bouman et al., 1999). Interaction among the farm types was taken into consideration by restricting the amount of hired labour on each farm type by the availability of off-farm labour from other farm types. Prices of some commodities within the region were also simulated in the model by taking into account the demand elasticity and regional supply of the commodity. They concluded that current production systems are unsustainable due to negative soil N and K balances. They then used the model to investigate various scenarios which included constraints on soil nutrient mining, and restrictions on the environmental impacts of biocide use, loss of N by leaching and greenhouse gas emissions. The results showed that reducing biocide use and N losses to the atmosphere could be achieved, but at the cost of reducing agricultural employment opportunities.

In a third case study in the region, Schipper et al. (2001) used SOLUS to evaluate policy issues of taxing biocides to reduce environmental contamination, and maintaining natural forests through payments of cash subsidies. Taxing all biocides at a fixed rate resulted in a reduction of biocide impact, but at a significant economic cost. However, a progressive tax where different rates were applied depending on the toxicity of the biocide resulted in an even greater reduction in biocide impact at only a small reduction in the economic surplus. To maintain natural forest, the Costa Rican government currently pays farmers US$40 ha^{-1} year^{-1} to not cut down trees. The results from the model showed that this value was too low to encourage farmers not to clear land, and that a value of more than US$120 ha^{-1} year^{-1} was needed to be effective. Increased wages were predicted to result in a move away from cropping to more land under pasture, as crops require more labour than does pasture.

The SYSNET project at the International Rice Research Institute (IRRI) is another example of the integration of crop models into land-use planning systems. SYSNET was a systems research network based at IRRI in the Philippines following on from the Simulation and Systems for Rice Production (SARP) project described in Chapter 12. The objectives of the project were to develop and evaluate methodologies and tools for land-use analysis, and apply them on a subnational scale to support agricultural and environmental policy making (Roetter *et al.*, 2000a). The project also involved scientists and stakeholders from India, Malaysia, Vietnam and the Philippines. A full description of the project and four case studies is given in Roetter *et al.* (2000a, b), but we thought it useful here to describe one of the case studies to illustrate the approach used.

In the state of Haryana in the Indo-Gangetic plains, rice and wheat, commonly grown in a double-cropping rotation, are the major food crops of the region. To meet the increasing demand of the growing population and to maximize farmers' and regional income, a recent policy goal is to double the food grain production in the next 10 years (Aggarwal *et al.*, 2000a). Regional stakeholders are also interested in finding optimal agricultural land-use plans that can meet this goal as well as maximize employment and income from agriculture, while minimizing pesticide residues, nutrient losses and groundwater withdrawal. The objective of the case study undertaken as part of the SYSNET project was to determine the magnitude of production possibilities, the associated environmental risks and the inputs required to attain the targeted production levels (Aggarwal *et al.*, 2000b). In the study, land units were defined by overlaying administrative boundaries on to agroecological zones based on soil and weather characteristics. Non-agricultural land (e.g. residential) was excluded. Databases relating to seasonal ground and surface water availability, labour, pesticides, fertilizers, transport facilities, and costs and prices of main farm inputs were developed on a district basis. Based on the current cropping pattern in different parts of the state, 14 land-use types (LUTs) were defined for the analysis. Three major dairy cattle types, cross-bred cows, buffaloes and local cows, were also included. Five technology yields were explored: potential yields, current yields and three levels (25, 50 and 75% of the difference) between these two extremes. Potential crop yields were estimated using the WTGROWS model (Aggarwal and Kalra, 1994) for wheat, the CERES model (Ritchie *et al.*, 1998) for rice and maize, and the WOFOST model (van Diepen *et al.*, 1988) for other crops. Current crop yields were obtained from state-level statistics, adjusted for the land-unit level to take into account factors such as salinity. Three goals were considered: maximizing food (i.e. rice + wheat) production, maximizing incomes and minimizing water use.

When the goal was to maximize food production, the results showed that, with no constraints, Haryana had the capability of producing nearly 40 million t of rice and wheat, about four times the current production. However, the capital requirement would be twice the current level, and water use would be three times as much as now. This analysis assumed that all farmers would use optimum technology; if this assumption was removed, the production level fell to 28 million t, still three times the current level. When water, capital and labour availability was also placed as constraints, the production fell to about 11 million t, only a little more than the current level. When the goal was to maximize farmer income but maintain food production at the same level as at present, the results showed that increased use of irrigation and N fertilizer and diversification into cash crops such as cotton, sugarcane and potatoes could increase incomes four-fold. When the goal was to minimize water use, the results showed that the current amount of food could be produced with almost half the irrigation water. This was due to the adoption of cropping systems such as fallow–wheat, which could produce food with higher water use efficiencies. The authors make the point that the approach as described only considered one level of stakeholders – the regional planners – and that other stakeholders such as farmers and village-level managers were not included.

6.2 Dynamic Simulation Approaches

Dynamic simulation models of processes (both biophysical and socio-economic) at the regional and national levels have been slower to be developed than optimization approaches, for reasons which include incomplete knowledge of all the processes involved (particularly those that are socio-economic in nature), the sheer complexity and number of processes occurring, and the amount of data needed for the construction and parameterization of such models. A major advantage of simulation approaches is their potential portability (Jansen and Schipper, 1995), in that they should be able to be used in different circumstances without rewriting the model, something that linear programming approaches suffer from.

Despite these constraints, some attempts have been made to use crop models directly for analysis at the regional level. Lal *et al.* (1993), for example, used the BEANGRO model to analyse regional productivity of bean at three sites in western Puerto Rico. Optimum cultivar selection, planting date and irrigation strategy were found to vary from one site to another. Chou and Chen (1995) discuss the combined use of satellite imagery, GIS and the CERES-RICE model for mapping potential land productivity and suitability under different irrigation management in China. An interesting approach is described by Thornton and Jones (1997), who developed a dynamic land-use model taking into account simple relationships for land quality (slope and drainage), distance from roads and markets, and gross margins from three crops (vegetables, maize, cotton) and used it to explore the evolution of land use over time.

The integration of crop simulation models with GISs is also emerging as a useful tool for planners. Beinroth *et al.* (1998) describe the development of AEGIS (Agricultural and Environmental Geographic Information System) for use with Decision Support System for Agrotechnology Transfer (DSSAT)-type models. They subsequently used the system to evaluate the feasibility of small irrigation projects for watersheds in the Andes of Colombia. A digital terrain model was used to classify the land within a catchment into three slope classes, with irrigation assumed to be feasible only on the 0–7% slope classification. The DSSAT dry bean crop model was used to estimate the response of crop production to irrigation inputs – dry bean was chosen as a typical crop with a short season and shallow root system. Estimates of river water flow were made at three positions in the catchments. The authors proposed an interactive compromise planning exercise in which stakeholders could define conditions and select fields for inclusion in the planned irrigation activities. Trade-offs between domestic requirements, irrigation demand and downstream use could then be calculated with AEGIS, which could be repeated over several years to get an idea of year-to-year variability in demand of each sector. Regular discussions during this process would allow new positions to be formulated and simulated in an iterative manner until a consensus was reached. The authors argue that such approaches of multi-party negotiation and consensus building are likely to become more important processes as conflicting goals of water use by different groups become increasingly contentious issues with increases in population.

In another example, Hansen *et al.* (1998) used AEGIS to assess the agricultural potential of previous sugarcane land in Puerto Rico: as a result of a depressed sugar market due to substitute sweeteners, sugar is no longer attractive for farmers to grow. Much of the land is now idle, although farmers are starting to grow vegetables in the valleys with substantial quantities of inorganic fertilizer and pesticides. Development of new types of agriculture is important for the economy, but could also pose risks to surface and groundwater quality. Agricultural planners need to have information on the likely impacts of alternative agricultural uses of these lands previously devoted to sugarcane production. AEGIS was used to investigate the trade-offs between production and the environmental impact of several crops including sugarcane, maize, bean, sorghum, tomato and soybean. They showed that tomato double-cropped with a cereal would increase profits for land use and would probably decrease risks of land erosion and N leaching, compared with a number of alternative cropping systems.

Similarly, Singh and Thornton (1992) used the CERES-RICE model to evaluate the effect of different N fertilizer rates on sorghum yields over 25 years in Maharashtra in India. Model output was presented as GIS maps showing the areas where maximum benefit could be achieved from fertilizer use, and also where the greatest risk (i.e. variability of yields) of using fertilizer were. The point was made that this information was useful at the government, industry and farmer levels; there was no evidence, however, that there was any uptake of the information by these sectors.

6.3 Limitations

Bouman *et al.* (1998) discussed the limitations of the optimization approaches using LP: the main one is that the input–output relationships generated by the technical coefficient generators represent equilibrium situations – where non-stable situations are generated, such systems have a limited life-span and the input–output combinations change over time. There is also a limit to the number of spatial units that can be handled in LP models, yet spatial variation is inherent at regional scale levels. Similarly, temporal variations in prices and yields are not accounted for well by such approaches.

Whichever approach is used, one of the major difficulties in application of systems analysis techniques for regional development planning is validation of the models used, which raises the question of whether we can be confident enough in their predictions to justify their use as guidelines for regional development and research planning. As the objective is to *explore* the options for regional development, rather than *predict* them, establishing data for model validation is generally impossible. Validation of individual components such as the crop models is possible, but this does not test for any errors introduced through linking them at a higher level. At higher levels of organization, interactions between system components become progressively more important and processes within components less important, so there is a need to ensure that these interactions represent reality. An attempt at validation of a model simulating various aspects of three widely contrasting socio-economic regions was made by van Keulen (1993). Despite the simplistic nature of the model he used, the results of the simulations reflected current development trends in similar areas in two of the regions, Israel and Egypt, with reasonable accuracy. Poorer agreement was obtained between the simulation and observations for the third region, in Australia, which was ascribed to differences in the main agricultural product, which was sheep in this case. However, it was difficult to know for sure in this example whether or not the model was giving the correct responses but for the wrong reasons.

6.4 Impact

With the exception of the work of Breman (1990) which was used to prioritize research issues for subsequent projects, and that of van Latesteijn (1993) which was used to formulate government policy, it is difficult to point to evidence of impact of any of these examples. The work in Costa Rica described by Bouman *et al.* (1999) was certainly aimed at policy makers, and was a response to a call from the government of that country for research to analyse the trade-offs between socio-economic and environmental goals for a range of policy options, but no evidence is presented of any change or refinement of policy as a result of the work.

Part of this must be due to the failure to include planners and policy makers in the overall research process. Van Keulen and Veeneklaas (1993) claim that the systems analysis approach allowed good interaction between scientists, planners and policy makers at workshops, but it is not stated whether this resulted in any uptake of the research results. Clearly, better dissemination pathways between research and policy making need to be established, although Spedding (1990) makes the point that policy makers are generally sceptical of systematic methods, 'as they are alarmed at the idea of it being publicly known where they are trying to get to, except in the most general terms, in case they never arrive!'

It may also be that the use of models at the policy-making level in developing countries has not gained acceptance, as there has not been a sufficient length of time for there to have been any impact as yet. In developed countries, models have been used in many explorative exercises in policy making (e.g. the Netherlands; R. Rabbinge, Wageningen, 2001, personal communication), and may eventually also come to be used more in developing countries as their utility becomes apparent.

Global Level Processes

7

Robin Matthews[1] and Reiner Wassmann[2]

[1]Institute of Water and Environment, Cranfield University, Silsoe, Bedfordshire MK45 4DT, UK; [2]Fraunhofer Institute for Atmospheric Environmental Research, Kreuzeckbahnstrasse 19, 82467 Garmisch-Partenkirchen, Germany

Although rapid industrialization over the last century has contributed to a major improvement in the living standards of millions of people, it has also brought with it serious problems. Industrial and agricultural emissions of carbon dioxide (CO_2), methane (CH_4), chlorofluorocarbons, nitrous oxide (N_2O) and other gases, mainly from the burning of fossil fuels, are resulting in an increase of these gases in the earth's atmosphere; CO_2 concentration, for example, is increasing at the rate of about 1.5 ppm year^{-1} (Keeling *et al.*, 1984). This increase in these so-called 'greenhouse gases' (GHGs) is contributing to a gradual warming of the planet by retaining more heat within the earth's atmosphere (Intergovernmental Panel on Climate Change, IPCC, 1996). General circulation models (GCMs), describing the dynamic processes in the earth's atmosphere, have been used extensively to provide potential climate change scenarios (Cohen, 1990). Based on current rates of increase it is generally accepted that CO_2 levels will reach double the current level by the end of the current century (i.e. 2100). According to current GCM predictions, a doubling of the current CO_2 level will bring about an increase in average global surface air temperatures of between about 1.5 and 4°C, with accompanying changes also in precipitation patterns (IPCC, 1996).

The relationship between climatic changes and agriculture is a particularly important issue, as the world's food production resources are already under pressure from a rapidly increasing population. Both land-use patterns and the productivity of crops are likely to be affected (Solomon and Leemans, 1990); it is vital, therefore, to obtain a good understanding, not only of the processes involved in producing changes in the climate, but also the effect of these changes on the growth and development of crops.

The emulation of future atmospheric conditions in field experiments is an intricate methodological problem and is generally limited to specific aspects of global change, i.e. increased CO_2 concentrations or ambient temperature (Ziska *et al.*, 1997). Thus the use of models appears to be the only feasible way of predicting the likely impact that climate change is likely to have in a wider context by taking into account the spatial and temporal variations of the different climatic parameters. Consequently, the literature on the use of crop models in climate change research is considerable. It is our aim in this chapter, therefore, to give just a few examples of how crop models have been used in this way, particularly in relation to rice. Rice is a particularly relevant crop in climate change research, as not only is its production affected by changes in climate, it also influences the processes of climate change through the production of CH_4 from decomposition of organic matter in the anaerobic soil conditions in which it is grown. It is also the second most important crop in the world after wheat in terms of production, and a staple for a large part of the world's population. However, it has been estimated that rice production must increase by 70% over the coming decades to meet the demands of a rapidly expanding population (IRRI, 1993). This is a major challenge for the agricultural research community and policy makers alike, particularly as yields in some of the more productive farmers' fields are already approaching the ceiling of average yields obtained on experimental stations (IRRI, 1993), thereby minimizing gains in production through improved management practices. The likely impact of climate change on rice production, therefore, only adds to an already complex problem, and is of paramount importance in planning strategies to meet the increased demands for rice over the next century.

Another global level process, the El Niño-Southern Oscillation (ENSO), is caused by changes in sea-surface temperatures in the eastern equatorial Pacific Ocean, but has also been shown to influence weather patterns in many other parts of the world. In this chapter we also consider examples where crop models have been used to predict responses to the ENSO, and how a knowledge of likely crop yields at different stages of its cycle could be used to benefit farmers.

7.1 Impact of Climate Change on Rice Production

A number of crop modelling studies on the likely effects of climate change on rice production have been carried out in recent years. Several of these have been limited to single countries or subregions within countries (e.g. Bangladesh (Karim *et al.*, 1991), Japan (Horie, 1988), China (Zhou, 1991)), and many have only considered the effect of temperature changes without including the influence of CO_2 (e.g. Okada, 1991). Others are based on statistical regression models only (e.g. Wang *et al.*, 1991). Yoshino *et al.*

(1988) predicted that lowland rice yields could increase in Japan by about 9% following a doubling of CO_2 and subsequent climatic changes as predicted by the Goddard Institute of Space Studies (GISS) general circulation model (Hansen *et al.*, 1988). Solomon and Leemans (1990), using a very simple model and long-term monthly-average climatic data in a worldwide study, predicted a yield increase of 0.4% for the current rice growing environments, but little change in the areas sown because of the sharp temperature and moisture gradient along the northern border of its primary distribution in eastern Asia.

Research on the effect of climate change on rice production began at the International Rice Research Institute (IRRI) in the late 1980s, when Penning de Vries *et al.* (1990) used the MACROS crop simulation model (Penning de Vries *et al.*, 1989) and weather data from four contrasting sites (the Netherlands, Israel, the Philippines and India) to simulate average grain yield, and its variability, of wheat and rice under both fully irrigated and rain-fed conditions. They used assumptions in their model of the way rice would respond to changes in temperature and CO_2 level based on what was known generally for a number of crops, summarized by Kimball (1983) and Cure and Acock (1986). Their results indicated that a doubling of the CO_2 level would increase yield by 10–15%, but that this would be offset by the effect of the expected accompanying rise in temperatures. These effects were the result of increased photosynthesis at higher CO_2 levels, and a reduced length of the growing season and increased maintenance respiration rates at higher temperatures. They also predicted that yield variability would be higher in cooler climates, particularly for rain-fed rice.

Jansen (1990) continued this analysis by using MACROS with historic weather data from seven sites in eastern Asia to investigate the possible impact of various climate change scenarios of increased CO_2 levels and temperatures on regional rice production. Simulations also included N-limited production, although this was incorporated simply by altering the rate of photosynthesis to one-third of its maximum rate. His results indicated that yields would rise if temperature increases were small, but would decline if temperatures increased more than 0.8°C per decade, with the greatest decline in crop yields occurring between the latitudes of 10 and 35°. In a subsequent study, Penning de Vries (1993) again used MACROS to evaluate the effect of temperature, CO_2 and solar radiation on rice yields in Asia. As with the previous studies, they found that increased CO_2 levels increased yield, but that there was a negative linear relation between temperature and yields due to its effect on photosynthesis, respiration and crop duration, such that the two factors more or less cancelled each other out. Based on these results, they concluded, somewhat sweepingly perhaps, that climate change would not be an issue in meeting the increased demand of 65% more rice predicted for the year 2020 (IRRI, 1989).

In the next stage of research, climate change scenarios were predicted by GCMs. Rosenzweig *et al.* (1993), working with collaborators from 22 countries, used a number of the International Benchmark Sites Network for Agrotechnology Transfer (IBSNAT) crop models to simulate likely changes in production of various crops under various GCM scenarios. They predicted that crop yields are likely to decline in the low-latitude regions, but could increase in the mid- and high latitudes. The different responses were related to the relationships between yields and temperatures that differ along a latitudinal gradient. At low latitudes, crops are currently grown near their limits of temperature tolerance, so that any warming subjects them to higher stress, whereas in many mid- and high-latitude areas, increased warming benefited crops currently limited by cold temperatures and short growing seasons. Again, only a limited number of sites (21 for rice in East and South-east Asia) were used. Leemans and Solomon (1993) used a simple model based on temperature, solar radiation and rainfall data in a geographic information system (GIS) environment to estimate the effects on the production of various crops on a global scale, and predicted an 11% increase in global production of rice. Their model did not, however, include physiological responses specific to rice, such as the response of spikelet fertility to high and low temperatures.

In a major study funded by the Environmental Protection Agency of the USA, Matthews *et al.* (1995a) used the ORYZA1 and SIMRIW crop simulation models to predict changes in rice production for all the major rice producing countries in Asia under three different climate change scenarios. These scenarios were those predicted for a doubled-CO_2 ($2 \times CO_2$) atmosphere by the General Fluid Dynamics Laboratory (GFDL), the GISS, and the United Kingdom Meteorological Office (UKMO) GCMs. In general, an increase in CO_2 level was found to increase yields, while increases in temperature reduced yields. Overall rice production in the region was predicted by the ORYZA1 model to change by +6.5%, –4.4% and –5.6% under the GFDL, GISS and UKMO $2 \times CO_2$ scenarios (see Table 7.1), respectively, while the corresponding changes predicted by the SIMRIW model were +4.2%, –10.4% and –12.8%. The average of these estimates would suggest that rice production in the Asian region may decline by 3.8% under the climate of the current century. Declines in yield were predicted under the GISS and UKMO scenarios for Thailand, Bangladesh, southern China, and western India, while increases were predicted for Indonesia, Malaysia and Taiwan, and parts of India and China.

These predicted changes were based on the assumption that cropping practices and genotypic characteristics are the same in the future as now. However, as Penning de Vries (1993) had noted, it is highly likely that, given the time scale involved, farmers and plant breeding programmes will develop new cultural practices and genotypes more closely adapted to the gradually changing conditions, thereby mitigating the negative, and enhancing the positive, effects of this change. Possible adaptations that may occur are the adjustments of planting dates to take advantage of longer growing seasons in northern climates or to avoid high temperature stress in hotter countries, and the use of varieties more tolerant to higher temperatures in the low-latitude regions. In rice, the fertility of spikelets is very sensitive to temperatures in the region of 33°C, where a difference of 1°C can result in a modest yield increase becoming a large yield decrease. Considerable variation between varieties in tolerance to high temperatures has been shown to exist (Satake and Yoshida, 1978).

Matthews *et al.* (1995b) explored these possible adaptations in some detail. In relation to changed planting dates, their results showed that at Beijing in northern China, the sowing window was predicted to widen considerably from 120 days to about 200 days under the UKMO scenario, so that both

Table 7.1. Estimated changes in total rice production predicted by the ORYZA1 model for each country and in the region under the three general circulation model (GCM) scenarios.

Country	AEZ	Current[1] production '000 t	GFDL Change (%)	'000 t	GISS Change (%)	'000 t	UKMO Change (%)	'000 t
Bangladesh	3	27,691	14.2	31,621	−5.0	26,298	−2.8	26,919
China	5	8,854	−7.4	8,201	0.3	8,881	−25.2	6,619
	6	79,872	0.8	80,484	−21.7	62,514	−19.5	64,334
	7	91,828	5.8	97,196	5.8	97,135	3.1	94,695
	8	2,361	−6.4	2,209	−14.2	2,026	−27.6	1,710
India	1	32,807	4.6	34,305	−10.8	29,272	−5.5	31,017
	2	49,949	1.8	50,849	−2.9	48,493	−7.9	46,002
	5	227	−7.4	210	0.3	228	−25.2	170
	6	26,628	5.4	28,069	3.2	27,480	−1.3	26,287
	8	1,011	−6.4	946	−14.2	867	−27.6	732
Indonesia	3	44,726	23.3	55,155	9.0	48,748	5.9	47,387
Japan	8	12,005	−6.4	11,231	−14.2	10,300	−27.6	8,696
Malaysia	3	1,744	24.6	2,173	17.6	2,050	26.8	2,211
Myanmar	2	13,807	21.5	16,776	−10.5	12,356	1.2	13,974
Philippines	3	9,459	14.1	10,797	−11.8	8,340	−4.7	9,018
South Korea	6	8,192	−13.6	7,078	−5.3	7,755	−21.9	6,401
Taiwan	7	2,798	11.8	3,128	12.8	3,156	28.0	3,583
Thailand	2	20,177	9.3	22,044	−4.7	19,230	−0.9	19,989
Total		434,136		462,472		415,129		409,743
Change (%)				+6.5		−4.4		−5.6

Current actual production ('000 t) in each AEZ on a country basis are adjusted by the simulated changes in total annual production (from Matthews *et al.*, 1995b).
AEZ, agroecological zone; GFDL, General Fluid Dynamics Laboratory; GISS, Goddard Institute of Space Studies; UKMO, United Kingdom Meteorological Office.
[1]Source: IRRI (1993).

earlier and later sowings become feasible. Although potential yields were lower than under the current climate due to the effect of increased temperatures on spikelet sterility, the wider sowing window allowed the possibility of two rice crops per year. Under the UKMO scenario, if the first crop was sown between the beginning of February and the end of March and the second in June/July, the total annual production was much higher than could be achieved with a single crop sown in early May under the current climate. However, the variability of yields was also higher, suggesting that while a warmer climate would potentially allow rice farmers in this area to move from the current single-cropping system to double cropping, the risks associated with this change would also be greater.

This analysis was extended to estimate the potential effect on China's national rice production brought about by a move from single cropping to double cropping in the areas where it becomes possible (Matthews *et al.*, 1995b). It was found that the sites of Nanjing, Chendu and Guiyang could support a transition to double cropping under all three GCM scenarios, and additionally Beijing under the UKMO scenario, as discussed above. All of these sites support only single rice-cropping systems at present (Defeng and Shaokai, 1995). Taking this transition into account gave predicted changes of +44%, +37% and +42% in overall national rice production for the GFDL, GISS and UKMO scenarios, respectively. These were considerably higher than the corresponding values of +12.2%, +2.0% and +5.6% calculated in the analysis of Defeng and Shaokai (1995) with no adjustment to the cropping system. Provided this transition occurs and that there are no other adverse factors emerging, it would seem that the predicted changes in climate will have a beneficial effect on Chinese rice production.

A similar analysis was carried out to investigate the effect of changing planting dates to avoid high temperature spikelet sterility in currently warm climates such as southern India (Matthews *et al.*, 1995b). At Madurai in the state of Tamil Nadu, for example, all three future scenarios were predicted to reduce rice yields considerably, with the ~3.5°C predicted increase taking temperatures above the critical value when spikelet sterility increases sharply. Delaying planting by 1 month to avoid these high temperatures in September, when flowering currently occurs, resulted in a large increase in yields again, although this delay meant that the following dry season crop would also be planted later, thereby taking its flowering time into a high-temperature period. Thus, while adjustment of planting dates may be able to maintain yields in the main planting season, often a second crop may not be attainable and the total annual production may fall.

To evaluate the effects of selecting for more temperature-tolerant genotypes, the incorporation of a +2°C increase in tolerance resulted in the yields rising even higher than the pre-scenario level for the Madurai site. Penning de Vries (1993) reasoned that future genotypes would be likely to be selected for tolerance of spikelet fertility to high temperatures, and therefore, that this factor did not need to be taken account of in estimating the effect of climate change on rice production. To test this, the ORYZA1 model was rerun with the same input data files as used for the regional analysis above (Table 7.1) but assuming no effect of temperature on spikelet sterility (Matthews *et al.*, 1995b). Rice production in the region was predicted to change by +14.9%, +15.6% and +12.9% under the GFDL, GISS and UKMO scenarios respectively. Within individual countries too, increases in yield were generally now predicted in most cases. Comparison of these results with those obtained in the first simulations indicated that the use of temperature-tolerant genotypes could more than offset the detrimental effect of increased temperatures under a changed climate.

These studies contain many uncertainties, partly due to the quality of the crop simulation models (Bachelet and Gay, 1993), partly from the use of limited sites for which historical weather data is available, and partly due to the reliability and resolution of climate predictions by the GCMs. For example, as most of the relationships relating the effect of temperature and CO_2 on plant processes are derived from experiments in which the crop's environment was changed for only part of the season, acclimation of the crop to changes in its environment is not generally taken account of in the models. Similarly, higher CO_2 levels have also been found to hasten development rates (Baker *et al.*, 1990), but, in current versions of both models, phenological development is taken to be independent of growth processes.

GCMs also have significant inherent limitations in projecting regional climate patterns (e.g. Houghton *et al.*, 1992; Bachelet *et al.*, 1993; King, 1993). The most significant limitations include: (i) poor spatial resolution; (ii) inadequate coupling of atmospheric and oceanic processes; (iii) poor simulation of cloud processes; and (iv) inadequate representation of the biosphere and its feedbacks. Moreover, the GCM predictions represent average changes that might be expected in the future following climate change. Changes in the frequency and intensity of hurricanes, the frequency of floods, and the intensity of the monsoons are much more important to the rice farmer than the average increase in monthly precipitation or temperature. Some of these issues are being addressed by the new generation of GCMs and the ability of modern computers to run these models in reasonable time frames. Modern GCMs with multiple nest capabilities can be 'zoomed' into selected grid cells for computing regional forecasts of future climates. As of now, this (potential) source of information has not yet been tapped for tropical agriculture. The trend in climate modelling towards regional forecasts may also accelerate the merger of crop, hydrology and socio-economic models into integrated landscape models. In the next step, these integrated models become feasible for larger regions; one example for this integrated approach is the ongoing WAVES project that assesses potential impact of climate change on the semiarid region of north-eastern Brazil (see http://www.pik-potsdam.de/cp/waves/).

Impact of these climate change studies is difficult to estimate, but the results from the rice work were included in the IPCC (1996) assessment of climate change impacts on agriculture (D. Olzyck, Oregon, 1998, personal communication) which contributed to policy making on climate change.

7.2 Greenhouse Gas Production

At the global level, CH_4 is the second most important greenhouse gas, and has been estimated to account for 15–20% of anthropogenic radiative forcing. Moreover, its concentration in the atmosphere has been rising in recent years. Rice agriculture is one of the major anthropogenic CH_4 sources, with current estimates ranging from 4 to 30% of the total anthropogenic contribution to the atmosphere (Houghton *et al.*, 1992). The increase in rice production required to meet the demands of an increased population has been estimated to increase CH_4 production by up to 50% (Bouwman, 1991). However, the IPCC has recommended immediate reductions of 15–20% in anthropogenic emissions of CH_4 to stabilize atmospheric concentrations at current levels (IPCC, 1990). The only feasible way in which these two opposing requirements can be met are by using crop management practices that reduce CH_4 emissions without affecting crop yields.

To address these issues, a multi-national project, coordinated by IRRI in collaboration with selected national agricultural research systems in major rice-growing countries of Asia, was established in 1993. The aims of the project were: (i) to provide more accurate estimates of CH_4 emission rates, and (ii) to develop strategies that would mitigate CH_4 emissions from rice fields without sacrificing crop yields. Experimental data on CH_4 emissions and the factors influencing them were collected from eight sites in five Asian countries, namely India, China, Indonesia, Thailand and the Philippines. An important part of the project was the use of this experimental data to develop a simulation model describing the processes involved both in CH_4 emission and in crop yield formation. Given the more advanced state of crop yield models, it was decided that the most efficient strategy to accomplish this goal was to integrate new subroutines describing the methane budget into a well-tested existing crop simulation model. The CERES-RICE model was chosen, as it has been relatively well tested in a range of environments (e.g. Bachelet *et al.*, 1993) and already has routines describing the main crop components involved in CH_4 dynamics, i.e. organic matter decomposition, root growth and death and root exudation. A module describing the steady-state concentrations of methane and oxygen in the soil (Arah and Kirk, 2000) was added. The development, validation, and use of the resulting MERES (Methane Emissions from Rice EcoSystems) model to upscale experimental field data to the national level and to evaluate potential mitigation strategies is described in a series of four papers (Knox *et al.*, 2000; Matthews *et al.*, 2000a, b, c).

MERES was used together with daily weather data, spatial soils data, and rice growing statistics to estimate the annual methane emissions from China, India, Indonesia, the Philippines and Thailand under various crop management scenarios. Four crop management scenarios were considered: (i) a 'baseline' scenario assuming no addition of organic amendments or field drainage during the growing season; (ii) addition of 3000 kg dry matter (DM) ha^{-1} of green-manure at the start of the season but no field drainage; (iii) no organic amendments, but drainage of the field for a 14-day period in the middle of the season and again at the end of the season; and (iv) addition of 3000 kg DM ha^{-1} of green-manure and field drainages in the middle and end of the season. The level of green-manure used was equivalent to the estimated national average use in China, and field drainage in the mid-season had been proposed as a means to reduce CH_4 emissions by introducing oxygen back into the soil. For each scenario, simulations were made at each location for irrigated and rain-fed rice ecosystems in the main rice growing season, and for irrigated rice in the second (or 'dry') season. Overall annual emissions for a province/district were calculated by multiplying the rates of CH_4 emission by the area of rice grown in each ecosystem and in each season.

Using the baseline scenario, annual CH_4 emissions for China, India, Indonesia, the Philippines and Thailand were calculated to be 3.73, 2.14, 1.65, 0.14 and 0.18 Tg CH_4 year^{-1}, respectively (Table 7.2). Addition of 3000 kg DM ha^{-1} green-manure at the start of the season increased emissions by an average of 128% across the five countries, with a range of 74–259%. Drainage of the field in the middle and at the end of the season reduced emissions by an average of 13% across the five countries, with a range of –10% to –39%. The combination of organic amendments

Table 7.2. Predicted annual methane emissions (Tg CH_4 year^{-1}) from each of the five countries in the study of Matthews *et al.* (2000c).

Country	Rice area (km²)	Scenario			
		1	2	3	4
China	323,910	3.73	8.64	3.35	7.22
India	424,947	2.14	4.99	1.88	4.07
Indonesia	110,088	1.65	2.87	1.00	1.90
Philippines	36,205	0.14	0.50	0.12	0.39
Thailand	96,442	0.18	0.42	0.14	0.32
Total	991,591	7.83	17.42	6.49	13.90
% Change from baseline			128	–13	86

Scenarios are (1) baseline scenario: continuous flooding and no organic amendments, (2) continuous flooding + 3000 kg DM ha^{-1} as green-manure, (3) field drainage and no organic amendments, (4) field drainage + 3000 kg DM ha^{-1} green-manure.

and field drainage resulted in an increase in emissions by an average of 86% across the five countries, with a range of 15–176%. The sum of CH_4 emissions from these five countries, comprising about 70% of the global rice area, ranged from 6.49 to 17.42 Tg CH_4 year^{-1} depending on the crop management scenario.

The limitations in this study were of two types – those arising from uncertainties in the model itself, and those related to the scarcity of the input data. In the case of the model, many of the relationships describing the behaviour of the processes involved in CH_4 emissions were derived from a limited number of experiments, some in laboratory conditions, and are therefore not fully tested, particularly for field conditions. For example, the rate of root exudation was based on one laboratory experiment. There was also considerable uncertainty in the root death rate – this was estimated as a constant 0.5% per day of the root biomass present, but little measured data exists to support this value. It was also assumed that the rate of substrate supply for the methanogens from fermentation was not a limiting factor (i.e. that all substrate available on a given day was consumed within that day). The transmissivity of the plants to gaseous movement of CH_4 and O_2 was also an estimate, and was assumed in the current model to remain constant throughout the season, although there is evidence to suggest that this is not the case. As far as the input data was concerned, major uncertainties were in the estimation of the initial oxidized alternate electron acceptor pool in the soil, and in the quantity of organic fertilizer applied to rice fields. The sparseness of weather data sites in some countries was also cause for some concern; large areas in both India and China, for example, were represented by only a few stations. All of these limitations, and their implications, are discussed in more detail in Matthews *et al.* (2000c).

Another model used to predict emission rates of trace gases is the DNDC (DeNitrification and DeComposition) model describing carbon and N biogeochemistry in agricultural ecosystems and forests (Li *et al.*, 1994; Stange *et al.*, 2000). The current version of the model involves 28 crops including rice, and requires input data on weather, soil and detailed descriptions of agricultural practices. The model can be run in a regional mode that directly incorporates GIS files on climate, soil properties, crop acreage, livestock population and map data. Model outputs include the temporal changes in a number of soil variables (given in vertical profiles), crop variables, as well as flux rates of CO_2, CH_4 and N_2O. The model has been used to predict N_2O emissions in the USA and in China (Li *et al.*, 1996, 2001) – predicted total emissions for China agreed closely with those calculated using the IPCC methodology, although geographical patterns deviated substantially (Li *et al.*, 2001). The model has been mainly developed from data obtained in the USA and northern China, which may, at this stage, affect its applicability to the tropics.

Efforts aimed at mitigating global climate change need to be seen against the background of economic development in tropical countries. While the rationale of the United Nations Framework Convention on Climate Change was to curtail GHG emissions, it also recognized the right to development. The ensuing Kyoto Protocol stipulated clear emission targets for developed countries, while the developing world agreed to broad cooperation on mitigation. Thus, irrespective of the fate of the Protocol, developing countries have to find their role in international climate policy. In terms of impact, therefore, the IRRI project described above on methane emissions from rice fields, to which the MERES model contributed, has assisted rice-producing countries in Asia in the following ways:

1. The overall project addressed the problem that many developing countries had in complying with the stipulations of the United Nations Framework Convention on Climate Change due to insufficient field data, knowhow and infrastructure. Baseline data was generated for major rice ecosystems for accounting for national inventories and exploring mitigation options, and identified a suite of mitigation management options that have no adverse impacts on rice productivity or the environment.

2. Many rice-growing countries were, directly or indirectly, blamed for major contributions to global warming due to claims of high methane emissions. The project demonstrated that rice production does not exert a major force on the greenhouse effect at the global scale, although it may be a major component to national GHG budgets of some Asian countries. The project identified crop management practices that can be modified to reduce emissions without affecting yields.

3. The project also identified ways in which CH_4 emissions could be reduced, taking into account various socio-economic factors. For example, (i) intermittent drainage in irrigated systems reduces emissions and can also save water; (ii) biogas technologies reduce fuel consumption and supply organic manure with lower emission potential; (iii) improved crop residue management through composting, mulching and early incorporation can also reduce emissions; and (iv) direct seeding results in less labour input and less methane output.

4. Rice-growing countries feared that political pressure from the developed world might adversely affect industrial development and constrain land-use options for resource poor farmers. However, the project showed that: (i) the use of chemical fertilizer in rice entails low CH_4 emissions, counterbalancing emissions during production and application; (ii) emissions from low-yielding rice ecosystems, i.e. rain-fed, upland and flood-prone rice, do not justify mitigation; (iii) irrigated rice systems with high baseline emissions offer win/win options for emissions and productivity; and (iv) high-emitting rice systems could attract 'clean development mechanisms' funded by industrialized countries.

7.3 The El Niño-Southern Oscillation

The El Niño Southern Oscillation (ENSO) refers to shifts in sea-surface temperatures in the eastern equatorial Pacific Ocean and related shifts in barometric pressure gradients and wind patterns in the tropical Pacific. Depending on the surface-temperature anomalies, ENSO activity may be warm (El Niño), neutral or cool (La Niña). During a strong El Niño episode, ocean temperatures can locally average 2–5°C above normal in the eastern tropical Pacific near the west coast of South America, sometimes resulting in nearly uniform surface temperatures across the entire equatorial Pacific. On the other hand, during a strong La Niña episode, ocean temperatures locally average 1–4°C below normal in the same region, resulting in large east–west variations in ocean temperatures across the equatorial Pacific. Even though the phenomenon occurs in the tropical Pacific, it also influences weather variability across much of the globe, and correlations between ENSO activity and agricultural productivity have been established for a number regions (e.g. Garnett and Khandekar, 1992). Because these correlations are reasonably reliable, various researchers have suggested that prior knowledge of ENSO events could be used to tailor agricultural production decisions to offset the negative impacts of unfavourable conditions or to take advantage of favourable conditions (e.g. Messina et al., 1999).

In South Africa, for example, Singels and Bezuidenhout (1998) found that the probability of low rainfall was increased dramatically during the warm El Niño phase. Singels and Potgieter (1997) evaluated the reliability of three drought predictors based on the ENSO by comparing predicted and actual drought seasons in the Glen region. The best correlation between observed and predicted values was obtained when a strongly negative Southern Oscillation Index (SOI) phase during November was used to forecast below normal seasonal rainfall, although predictions of agricultural drought were less reliable. They then used the PUTU maize model to simulate the yields and gross margins obtained under a standard production strategy and that of different drought mitigation strategies such as reduced plant population or N fertilizer application rates. The results predicted that not sowing at all during drought seasons increased the simulated mean gross margin by 11%. Other production strategies were less efficient. A similar study for sugarcane, in which long-term monthly rainfall totals were analysed for five regions and the CANEGRO model was used to simulate yields for each region, showed that a poor distribution of rainfall led to low simulated yields (Singels and Bezuidenhout, 1998). However, the relationship between El Niño events and actual historical yields obtained from sugar industry records was confounded by other factors (Singels and Bezuidenhout, 1999). Using the model to exclude non-climatic factors indicated that cane yields are reduced by significant margins following seven out of nine El Niño events. Results showed that the phase of the SOI during spring is a reasonably reliable indicator of low rainfall during the subsequent midsummer and of low yields during the subsequent milling season (Singels and Bezuidenhout, 1999).

In Zimbabwe, Phillips *et al.* (1998) analysed climate data from four sites in four of the five agroecological zones (AEZ) in the country with respect to the different ENSO phases. They then used the CERES-MAIZE model parameterized for soil conditions typical of each area to predict yields under two N fertilizer treatments and three sowing dates. As in South Africa, at all four sites there was a decrease in seasonal rainfall associated with the El Niño phase compared with the neutral and La Niña years. While average simulated maize yields were generally lowest in the El Niño years, there was still a large variation in rainfall patterns and yields at each site within each ENSO phase, indicating that more precise seasonal rainfall predictions would be necessary if forecasts were to be of any use in practical crop management decision making. However, the results did show the potential of knowing in advance when favourable cropping seasons as opposed to poor ones were likely to occur, particularly in respect to more appropriate N management for the marginal sites.

In the Argentinean Pampas, the ENSO accounts for a large part of the inter annual variability of the climate. During the colder La Niña phases, maximum temperatures and solar radiation are usually higher, and minimum temperature and rainfall lower. The opposite is usually observed with the warmer El Niño events. There is a corresponding influence on crop production – maize, soybean and sorghum yields tend to be lower during the cold phases, whereas sunflower yields tend to be marginally higher. In the warm phases, crop yields are generally higher, with maize responding the most positively to the associated increases in rainfall. As part of a multi-institutional, interdisciplinary research programme established to study these effects in detail, Messina *et al.* (1999) developed a non-linear optimization model to explore the potential for tailoring land allocation among crops at the farm scale to the different ENSO phases for two regions of the Pampas. Using the CERES model for maize and wheat, the CROPGRO model for soybean, and OILCROP-SUN for sunflower, they estimated distributions of crop yields due to historic climate variability for given soil and genotype parameters, initial conditions, and crop management scenarios. At the end of each year, the model decided what crop to grow in the following year by maximizing the expected utility of wealth by selecting the cropping system with the highest mean net return. It was assumed that prices were known at the decision time, and that they were fixed (e.g. by contract) for the following year. Four levels of risk aversion were considered and three levels of initial household wealth. For moderate risk aversion, the results predicted allocation of land to crops similar to that currently observed in the two regions. Results showed that farm income could be increased by 9% on average and up to 20% if optimum crop combinations were selected for the different ENSO phases. Extrapolated to the 11 Mha of the whole Pampas region, this represented

US$166 million per year (Jones *et al.*, 2000). It was concluded that the model had potential as a farm decision-support tool, but it is not stated if there has been any uptake of the results.

Similar analyses were made in Australia for wheat production (Hammer *et al.*, 1996b) and groundnut production (Meinke *et al.*, 1996; Meinke and Hammer, 1997). Using a system of classifying the ENSO cycle into five different phases based on the value and rate of change in the SOI, Hammer *et al.* (1996b) demonstrated up to 20% increase in profit and/or 35% reduction in risk if N fertilizer and choice of cultivar were tactically adjusted in response to knowledge of the forthcoming rainfall and frost occurrence. Similarly, for groundnut, Meinke and Hammer (1997) showed that in years when the November–December SOI phase is positive there is an 80% chance of exceeding average district yields, while in years when the November–December SOI phase is either negative or rapidly falling there is only a 5% chance of exceeding average district yields, but a 95% chance of below average yields. It was suggested that this information could allow the groundnut processing and marketing industry to adjust strategically for the expected volume of production. It is not stated, however, whether this has occurred.

Part 2
Models as decision-support tools

Decision Theory and Decision Support Systems

8

William Stephens

Institute of Water and Environment, Cranfield University,
Silsoe, Bedfordshire MK45 4DT, UK

8.1 Decisions, Decisions, Decisions

A decision implies a clear-cut resolution of a problem, yet the process of reaching the decision can be anything but straightforward. Any number of tangible and intangible factors must be taken into account with variable amounts of data, intuition and experience all helping to inform the decision maker. The more complicated the system, the more we tend to rely on intuition and experience.

Agricultural enterprises are generally highly complex systems since farmers' activities are strongly affected by the external environment as well as by their own goals and culture. In managing these activities they are faced by an enormous range of decisions throughout the year. Some are automatically dealt with but many require them to make a conscious judgement on the action to be taken. Within the context of agriculture, decision support systems (DSSs) are, as their name implies, designed to help farmers take decisions by evaluating outcomes of alternative actions in the light of available information. In reviewing the success or otherwise of DSSs designed for agriculture it is first useful to discuss briefly the characteristics of a decision and then examine types of decision-making situations.

8.2 Characteristics of Decision Making

By definition, there must be a problem before a decision can be taken. Entire careers have been spent on researching problem definition and decision analysis across the whole gamut of human activity, but in the context of this book on the application of crop–soil simulation models in developing countries, a brief introduction must suffice.

Decision makers typically face many problems at any given time so that they may have multiple objectives relating not only to the immediate problem but also to various other aspects of their work and life (Goodwin and Wright, 1998). The possible decisions are confounded by uncertainty and by the decision maker's attitude to risk. The difficulty of problem definition is exacerbated by the complexity of the situation and, in many cases, will also need to be agreed by multiple stakeholders (Goodwin and Wright, 1998).

Problem definition may start at a fairly abstract level with a vague sense that something is wrong. By gathering qualitative and quantitative information a 'rich picture' (Checkland, 1981) can be developed. In this context, Ackoff (1981) describes decision makers as operating in a 'mess', as they are confronted by a number of interacting and overlapping problems. Individuals generally deal with this complexity using a broad conceptual model representing their own view of the world, or *weltanschauung* (Checkland, 1981). One result of this process is cognitive inertia – the inability to apply simpler solutions to problems because of prior experience (Goodwin and Wright, 1998).

Once the 'rich picture' has been analysed to identify and agree key issues, these can be addressed individually. It is this process of problem definition that seems to have been inadequate in many DSSs that have been developed. Problem definition requires the identification of an objective and of the obstacle(s) that prevent the objective from being achieved. How one tackles a problem will depend on how it is defined (Finlay, 1989) and this will depend on who defines it. Tackling a problem then becomes an exercise in solving, resolving or dissolving the obstacle to achieve the objective (Ackoff, 1981). Here, solving means achieving the optimum solution whereas resolving seeks only an answer that is good enough. Dissolving removes the obstacles by changing the conditions that caused the problem in the first place.

Decision-making strategies used by individuals vary with the amount of information they are presented with and with the sequence in which they obtain it. The process is also affected by the amount of pressure that they are under to reach a decision, for example due to time or importance of the outcome. One approach, known as 'elimination by aspects', focuses on identifying the most important attribute and a cut-off value below or above which alternatives will not be considered. When buying a computer, for example, the price may be considered to be the most important attribute. Once a price band has been established, then the purchaser may evaluate the alternatives in terms of their processing power, the size of the hard disk or the amount of RAM. If there are still comparable alternatives, then the perceived reliability of the manufacturer may be used to discriminate between them. The final decision on the type and model to buy can be affected by the sequence of the aspects selected and the non-compensatory nature of the strategy. For instance, a computer eliminated as being just outside the desired price range may have benefits in terms of reliability and performance that more than compensate for the extra cost.

When faced with sequential rather than simultaneous decisions then an individual may identify a set of aspirations against which to judge the alternatives. They will then form an opinion on the suitability of each alternative in turn until they find a satisfactory option. This may not be the optimal solution but meets sufficient of their aspirations to be acceptable. An example of this approach could be the purchase of livestock at an auction where the farmer may decide to purchase animals that do not meet all his requirements, as either there are none that do, or else he feels that the probability of being successful within the desired price range is too low due to competition with other potential purchasers. This approach, rather inelegantly termed 'satisficing' (Simon, 1979), can also be applied to simultaneous choices where the number of alternatives is too large to consider in detail.

The disadvantage of these intuitive approaches to decision making, according to decision analysis theorists and practitioners, lies in their inability to apply trade-offs and in the potential irrationality of the solutions (Goodwin and Wright, 1998). Thus in Ackoff's (1981) terms, they tend to involve resolving rather than solving problems. Whether this is actually a major disadvantage depends on your own point of view.

The simplistic examples presented above illustrate some approaches to decision making that operations research (OR) has sought to 'improve' through more rational and systematic methods. However, Schon (1983) warned that 'complexity, instability and uncertainty are not removed or resolved by applying specialised knowledge to ill-defined tasks'. Notably, the success of OR in highly structured environments within the military and industry has not been repeated in the more dynamic and 'messy' field of business management (McCown, 2001).

Schon (1983) has suggested that the response of modellers to the rather unpalatable discovery that modelling was not a panacea for all decision making was, by and large, to carry on regardless. Almost two decades later, the situation appears not to have changed substantially with a plethora of process-based and empirical models being produced, few of which have had any impact on farmer practice.

8.3 Definitions of Decision Support Systems

Several definitions of DSSs have been reviewed by Finlay (1989). Early definitions of DSSs sought to differentiate them from management information systems (MISs) and 'management science' or OR. Keen and

Scott-Morton (1978) suggested that DSSs have most impact on decisions where there is sufficient structure for computer and analytical aids to be useful but where managers need to use their judgement. The main benefit of successful DSSs has been to extend the range and capability of the decision processes, thus improving individual effectiveness. Finally, the relevance of DSSs is in the creation of a supportive tool that is under the control of the decision maker and does not make the decision for them by automating the decision process, pre-defining objectives or imposing solutions (Keen and Scott-Morton, 1978). In our view, these characteristics tend to define what makes a good DSS in that it is much less likely to be useful if it does not conform to these criteria.

The term DSS covers a wide range of systems, which vary greatly in their structure and complexity. DSSs have evolved over the years from rudimentary single decision rules to multiple-criteria optimization software. In their simplest form, a decision support tool can be a pest management threshold calculated using empirical relations and field data on a calculator. In a sophisticated form they can be interactive computer systems that utilize simulation models, databases and decision algorithms in an integrative manner. DSSs, in whatever form, should produce decision rules for intervening in a system directly or indirectly (Teng *et al.*, 1998). Newman *et al.* (2000) define DSSs as computer systems that assist the user in complex problem solving or decision making. DSSs typically have quantitative output and place emphasis on the end-user for final problem solving and decision making.

In terms of Ackoff's (1981) three problem-solving approaches, many DSSs aim at solving problems, but, in doing so, may have much greater information requirements than an approach that concentrates on problem resolution. If the obstacle can be dissolved then there is no need for a DSS to be developed (Finlay, 1989).

It is worth considering here where DSSs fit in the wide range of different model types. In a useful book on decision support systems for grazing lands (Stuth and Lyons, 1993), Whittaker (1993) proposed three spectra to characterize models (Fig. 8.1). In the first spectrum, different modelling approaches can be placed on the continuum between perception and reality. Whittaker asserted that 'human beings constantly strive to move toward the left end of this spectrum' (Fig. 8.1a). This may be true of the objective, reductionist section of the scientific community but the universality of this statement is questionable. Leaving this aside, the range of approaches from physical model to expert system shows the increasing abstraction as one moves from exact replica to a subjective representation based on expert opinion.

The information resulting from the model system illustrates the data quantity and value spectrum ranging from occurrences and observations through information and knowledge to decisions, recommendations and implications (Fig. 8.1b). Increasing time and effort is required to distil and interpret the available information as one moves from right to left, yet this

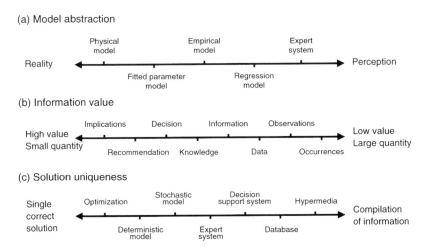

Fig. 8.1. Spectra of (a) model abstraction; (b) information value; and (c) solution uniqueness (redrawn from Whittaker, 1993).

is often desirable since it is easier to deal with small quantities of high-value information.

The third spectrum places DSSs in the context of other forms of modelling as being closer to a system for compiling information than for achieving a unique solution (Fig. 8.1c). Whittaker again asserted that humans strive towards achieving a single 'correct' solution, though he did admit that this compromises the holistic viewpoint. This desire for a single solution appears to be one of the key shortcomings of OR and decision-making approaches, as the perception of what that optimal solution is will depend on how the problem is framed and on the local context in which it is posed.

As part of the same publication, Ison (1993) explored the difference between 'hard' and 'soft' systems and drew the distinction between hard systems approaches to optimizing a solution compared with a soft systems learning methodology. He argued that while computer-based DSSs may have a place, they would only be successful if the end users are able to participate in the analysis and design of the system. He concluded that 'there exists a view that computers are but one means, rather than an "end", and the only means to decision support' (Ison, 1993).

In a challenging review of the gap between science-based decision support and the practice of farming, McCown (2001) develops this viewpoint and highlights the overriding importance of the local context, particularly the social context in which farmers practise, in determining the likely uptake of models and decision support systems. McCown's analysis takes a different approach from Whittaker's (1993) in that while the types of systems in the left hand column of Table 8.1 are present on Whittaker's spectrum of model abstraction, they are placed in chronological order of development. In direct contrast with Whittaker's assertions, McCown's view

Table 8.1. Comparison of seven types of analysis/intervention (McCown, 2001).

Type of systems analysis/ intervention	Characteristics of systems analysis/ intervention	Model of production system	Model of management system
Empirical study of the farm as a business system	Collaborative development of farm recording systems; use of average costs to aid planning	Records of production	Records of costs
Marginal analysis using production economics theory	Recommended action based on whole farm resource optimization; analysis included marginal returns and opportunity costs	Technical coefficients provided by production functions; static input–output transformations	Notionally specified decision problem; notionally specified conditions for economic model
Dynamic simulation of production processes	Recommendations based on single factor simulations	Dynamic model of production processes; notional initial conditions for simulation	
Decision analysis using dynamic simulation of production processes	Enhanced recommendations based on optimization of production inputs	Dynamic model of production processes; notional initial conditions for simulation	Notionally specified decision problem; notionally specified conditions for economic model
Decision support system	Interactive decision support system on farmer's computer	Simple model or abstracted output of a complex model of production categories of conditions	Notional decision problems
Expert system	Interactive expert system on farmer's computer	Table of action outcomes	'If . . . then . . .' model of expert manager's procedures
Cooperative learning using simulation-aided problem discussion	Facilitation of learning aided by customized simulation response to farmer's felt problems and situation	Dynamic model of production processes; measured initial simulation	Participating farmer represents management system in setting of problem, customizing simulation and interpreting output

is that to effectively bridge the gap between decision support and farming there needs to be a definite step towards the cooperative learning approaches with a move away from single solution 'reality'.

The empirical study of farms as business systems began in the 1950s when economists perceived a dichotomy between farm management and technology. Then, in the 1960s, economic optimization became the prime basis for interventions, firstly with static production functions, and then using dynamic simulations of production processes. McCown (2001) identified a major discontinuity between decision analysis using dynamic simulation of production processes and DSSs and expert systems as the latter 'represent personalised, interactive aids to decision making'.

There is no clear-cut line between DSS and models used in research to provide information which may be used later to inform decision making. In the literature, many models used in research are promoted in terms of their ability to aid in decision making (e.g. agricultural strategies in the face of global warming, selection of optimum land management practices). This chapter therefore deals not only with specifically designed DSSs but also with models whose output is intended for use in practical decision support (as opposed to increased learning about plant or systems functioning for its own sake). Expert systems (ES) are another computer tool for decision making, using qualitative rather than quantitative reasoning. The rules of thumb on which they are based are sometimes formulated with the use of crop simulation models. Newman *et al.* (2000) suggest that DSSs are more widely used and known in agriculture than ES. This review in this book does not attempt to cover ES since they rarely involve the use of crop–soil simulation models.

8.4 Spatial and Temporal Scale

Decision making occurs at a variety of spatial and temporal scales or levels. These are often interdependent as different spatial levels often imply different time frames. Decision support on a larger scale (e.g. an agroecological zone) may imply greater system complexity though this will depend on the number of organizational levels that contribute to the problem. Decision-making based on a systems approach needs to consider the interaction between the many subsystems that make up the whole, in particular, agricultural, ecological and socio-economic systems (Dent and Thornton, 1988). Three examples of schemes for classifying the hierarchies of levels within different subsystems are presented in Table 8.2. In practice, most DSSs tend to fall into one or another subsystem and are confined to one or, at most, two levels.

Temporal issues also influence the decision-making process. Whisler *et al.* (1986) suggest that farmers make decisions in two general time frames – daily and annually. Daily decisions might cover the amount and/or timing

Table 8.2. Hierarchies of scales within ecological, agricultural and socio-economic systems (Hess and Stephens, 1998).

Ecological system	Agricultural system	Socio-economic system
Ecozone	Ecozone	Region
Basin	District	Community
Catchment	Farm	Village
Hillside	Field	Family
Patch	Plant	Individual

of agronomic operations such as irrigation, herbicides, pesticides, fertilizers or the timing of cultivation or harvest. These decisions tend to be based on rules-of-thumb, custom or opinion. Annual decisions include those such as which variety or area to plant or which machinery to purchase.

Penning de Vries (1990) developed a broader temporal categorization of on-farm management decisions:

- *Operational decisions* – choices during a cropping season (e.g. irrigation dates, amount of fertilizer, timing of insecticide spraying). Impact on crop: 1–25 days.
- *Tactical decisions* – choices made once per season (e.g. crop species, date of sowing, yield targets). Impact on crop: 5–50 weeks.
- *Strategic decisions* – impacts over several seasons (e.g. machinery, field/infrastructure improvement, education and training). Impact on crop: 0.5–10 years.

Although all types of decision (operational, tactical and strategic) may be applied at any spatial level, decisions taken at larger spatial scales (e.g. change in practices to counter the effects of global warming) tend to be strategic in nature while farm-level decisions are often operational or tactical. Considering spatial and temporal scale provides a framework in which to analyse the application of models. It also has implications in terms of costs and benefits and the time scale over which impacts can be expected. The person who makes the decision will also be determined by the scale of the problem. Regional strategic decisions are likely to be taken by government policy makers who may have very different approaches to small-scale farmers. For the purposes of this review, operational and tactical (action) decisions have been grouped to reduce complexity, and are compared with strategic (planning) decisions.

8.5 Application

In the following chapters, we present some examples of crop–soil models developed to support decision making. The models are broadly grouped according to the type of decisions they support (i.e. operational or strategic).

Not all the models mentioned show evidence of practical application. However, where they show potential they have been included to support the discussion. The extent to which a given model is discussed depends greatly on the amount of documentation available. The review is by no means exhaustive in term of models, but represents types of application that stand out in the literature. Many of the examples come from developed countries, since there is little evidence of DSS use in developing countries. However, these cases may provide insight into the potential role of DSSs in less-developed countries. In Chapter 11, we attempt to answer the question: why has the uptake of DSSs been so poor?

Tools to Support Operational Decision Making

9

William Stephens and Tabitha Middleton

Institute of Water and Environment, Cranfield University, Silsoe, Bedfordshire MK45 4DT, UK

Operational decisions have a short timeframe and relate to actions to be taken. They may be made once during a season (e.g. when to plant) or on a day-to-day or week-by-week basis (e.g. irrigation scheduling, when to apply fertilizer or the best dates of pest spraying). They are most likely to occur at a farm level. Penning de Vries (1990) suggests that crop modelling can best support tactical decisions by developing guidelines, rules, tables and maps. The following sections describe applications which the authors consider to be tactical. Areas such as on-farm pest management and irrigation scheduling show the most evidence of practical model application.

9.1 Pest Management

Teng and Savary (1992) define pest management as a set of activities in agricultural production aimed at keeping pest populations or injury within economically and socially acceptable loss levels. Crop models have been relatively successful in helping farmers to decide when to spray their crops and how much pesticide to use. However, their measurable adoption by farmers has still been low, when compared with the expectations of those who developed them. Published information suggests that their use has been greater in developed countries, where growers not only need to consider how to control pests but also how pesticide applications fit in with health, safety and environmental legislation. In developing countries, computer-based pest management is still a long way from realization, even in the more developed countries of Asia (Teng and Savary, 1992).

Simplified decision rules from pest–crop models and simplified pest models with economic values assigned to their outputs have been used for managing sugarbeet leafspot (Shane *et al.*, 1985), sweetcorn common rust (Teng, 1987), wheat diseases (Zadoks, 1981) and rice blast (Surin *et al.*, 1991). Detailed simulation models have been used to design strategies for insecticide use (Heong, 1990) and predict disease epidemics (Teng *et al.*, 1978). However, most of this work was within the context of research projects.

SIRATAC and EPIPRE are commonly cited examples of successfully applied operational decision support systems (DSS), and have provided a benchmark against which other DSSs have been judged (Hamilton *et al.*, 1991; Cox, 1996; Sinclair and Seligman, 1996). Although both were applied in developed countries, they may provide pointers as to the role that DSSs can play in developing countries. Both are on-farm pest management systems used by farmers in the developed world which share many features both in terms of design and application. Each system required growers to pay for membership and to supply field observations to a central processing centre, where simulations were used to provide growers with updated pest management recommendations on a field by field basis.

SIRATAC was a dial-up crop management system developed in Australia to assist cotton growers in making good tactical decisions about the use of insecticides in irrigated cotton on a day-to-day basis (Macadam *et al.*, 1990). It was run by SIRATAC Ltd, a non-profit commercial company formed in 1981 to market the system to the cotton industry. SIRATAC tried to reduce the risk associated with pesticide use by adopting the principles of integrated pest management (IPM). It consisted of several simulation models and a decision model helping the grower to decide whether or not to spray pesticide and which pesticide to spray. The area managed using SIRATAC increased steadily during the early 1980s. In 1981–1982 there was demand for a tenfold expansion (Cox, 1996). However, by 1985 it had reached a ceiling in adoption at 25% (by area) of the industry and its use declined after 1987 (Hamilton *et al.*, 1991). In 1989, SIRATAC Ltd went into voluntary liquidation, despite the fact that the area managed by SIR-ATAC was at a historical high, due to a predicted declining market share and cash flow problems (Cox, 1996). They were replaced by an informal user group (SUG) which continued the programme for a few more seasons, but by 1993 the group had ceased operation and the field support system was no longer available. A moratorium on SIRATAC development meant that only minor changes were made to the programme. Attempts to achieve similar functionality on a microcomputer failed.

The University of Western Sydney carried out an in-depth participatory analysis of the limited adoption of SIRATAC (Macadam *et al.*, 1990) and found many of the reasons for poor adoption were based on organizational issues rather than problems with the model itself.

The important themes that emerged from their survey were:

- Widely held negative views of non-users against SIRATAC and SIRATAC Ltd.
- Many growers did not consider SIRATAC's laudable but intangible benefits to be worth paying for. They wanted to see tangible cost saving benefits.
- Many people based their opinions on hearsay, and did not really know what SIRATAC was or what it could do.
- Many felt that they were often overriding the system and so abandoned it. They felt uncomfortable with the recommendations of the system (e.g. expensive sprays when cheaper ones would do, not enough spraying – they felt it may put crops at risk).
- Practical limitations, no office, computer or reliable telephone service.
- Growers felt threatened – they saw it as a man versus machine issue and felt that their own experience and knowledge was being undervalued.

EPIPRE (EPIdemic PREvention) was developed between 1977 and 1981 as a system for supervised control of diseases and pests in winter wheat. The EPIPRE project was an experiment in modern extension. It was considered as a development rather than a research project since it did not strive for new knowledge but rather the application of existing knowledge from areas of crop husbandry, phytopathology, entomology and computer science (Zadoks, 1981).

EPIPRE was intended to reduce the extent to which farmers had to rely on external advice (e.g. from chemical companies). One strength of the system was that the farmers were able to learn as they went along and were always aware of how the system worked. Farmers were recruited from Wheat Study Clubs and tended to be advanced, eager and willing to learn. They were trained in recognition of disease symptoms. The participating farmers carried out their own disease and pest monitoring and sent their field observations to a central team who entered them on a daily basis into a computerized data bank. The system produced recommendations for treatment optimizing financial returns for crop protection. There were three major decision options: 'treat', 'don't treat' and 'make another field observation'. When the season was over, every farmer received a record of his actions and the recommendations, as well as a financial account of the crop protection activities. The results were discussed with farmers in regional meetings. The farmers listened critically and often expressed appreciation of their improved expertise. Eighty five per cent of farmers participating in one year participated in the next. Complete adherence varied between 20 and 80%. Partial adherence was far greater than full adherence (Zadoks, 1981).

In 1981, 6% of Dutch winter wheat was covered by the system (Rabbinge and Rijsdijk, 1983) saving on average £15 ha^{-1} (Zadoks, 1981). EPIPRE was also implemented in Switzerland and Belgium. In 1982, the system was introduced on an experimental basis in England, Sweden and France

(Rabbinge and Rijsdijk, 1983). In Belgium the system functions under the auspices of the National Soil Service, and in 1992, EPIPRE was used to advise on disease control in 500 fields in Belgium and northern France (Smeets *et al.*, 1992). The research in Sweden (from 1982 to 1985) found that although EPIPRE was an interesting and useful system, it would require alteration if it were to be introduced to farmers for routine use. There was a need to incorporate meteorological parameters into the model and adjust the plant growth model to suit Swedish conditions as the model was still recommending unnecessary spraying. This extra work was considered to be worthwhile (Djurle, 1988).

Experience in Switzerland found that EPIPRE-treated fields yielded 3% less than traditionally treated fields (0.5% with corrected gross return) but that farmers were able to reduce the spray frequency by 20–100%. The conclusion was that although the farmers did not make more profit, the new practice was ecologically beneficial and helped to reduce selection pressure for pesticide resistance (Forrer, 1988), which can have disastrous results (Zadoks, 1981). In 1981, if all Dutch winter wheat had been treated with the EPIPRE recommendations it would have reduced the pollution load by 12 t (Zadoks, 1981).

In both cases, there was an initial steady increase in membership, which resulted in an improvement in pest management. However, both organizations then experienced a decline in membership as the growers felt that they had learned what the model would predict, and therefore did not need them any more (Sinclair and Seligman, 1996). Both models were a success in that they contributed to an improved level of pest management by developing farmers' understanding and by helping them to interpret their own field observations more effectively.

More recently, Murali *et al.* (1999) have claimed that PC-Plant Protection is the most widely used PC-based farm-level DSS for pest control in Europe with over 2000 copies sold by 1999. The relative success of this DSS stems largely from a concerted promotional programme through a series of local and national advertisements, television coverage and a price cut from US$350 to about US$100. The decisive factor appeared to be the price cut in 1995, after which about 1400 copies were sold. Since then, sales have averaged only 200 per year despite the intense marketing including a targeted mail shot to 22,000 farmers. The authors remark that the vision leading to the development of the system was the strategic goal of meeting the target reductions in pesticide usage in the national Pesticide Action Plan. However, the main reason for partially achieving the target was that high-dosage pesticides were banned or greatly restricted. This apparent success therefore appears to follow the same pattern as the earlier examples in generating early interest followed by a disappointing response thereafter.

Our requests for information from the AGMODELS listserver (an Internet discussion group concerned with agricultural models at agmodelsl@ unl.edu) revealed some undocumented examples of models used to help decision making. For example, Growth Stage Consulting Inc., based in Calgary, Canada, has been using growing degree-days models to predict crop development 'for many years' (Dr Meng Xu, Manager of Crop Model Development, 2000, personal communication). It uses these models to help farmers to control weeds by applying chemicals at the right crop stages. The company has expanded the crop model application from providing services with fax reports to an internet service. Users, including agriculture retailers, farmers, researchers and educators can log on to their website at http://www.growthstage.com to obtain their services. These are currently limited to the USA and Canada due to local weather data resources.

Another example, though one which does not strictly fall within the remit of this review, is WORMBASE, a system developed by a team at Imperial College London, for army worm forecasting in East Africa. This system has been installed in Kenya and Tanzania and is currently being used for operational purposes (Jonathan Knight and Geoffrey Norton, University of Queensland, 2000, personal communication). In neither of these two examples did the correspondents indicate the scale of uptake or impact.

Knight (1997) suggests that DSSs have a great potential role in the transfer of new information from researchers investigating IPM and ways to reduce the negative environmental impact of pest control to farmers. He claims that this transfer is currently weak due to the great volume of information being produced. DSSs would be one way to integrate the many important new research outputs. Doyle (1997) supports this view, and recommends that, in the case of biological and genetic control, modelling the consequences of interventions is imperative, given some of the potential risks.

In contrast, Way and van Emden (2000), in their review of IPM in practice, take the opposite view and ask rhetorically whether, despite the very large quantity of literature, biological pest models have ever provided insights or identified knowledge gaps that were not evident from original research. The answer, based on a number of reviews (e.g. Barlow, 1998) is, they feel, a resounding 'no'. This leads on to an assertion that too much time and funding has been devoted to strategic modelling in IPM rather than to relevant field research (Way and van Emden, 2000).

9.2 Irrigation Scheduling

Irrigation scheduling is the process used by the irrigator to decide when to irrigate the crops and how much water to apply. It is a means of optimizing agricultural production while conserving water (van Hofwegen, 1996). Criteria for irrigation scheduling can be established by several approaches based on soil water measurements, soil water balance estimates and plant stress indicators in combination with simple rules or sophisticated models (Itier, 1996). Cabelguenne (1996) claims that there are at least 140 models based on the use of Doorenbos and Kassam's (1979) water production functions, but that such models are unable to forecast correctly the effect of water constraint on the growth of the plant since they take no account of dynamic functions. Mechanistic agronomic models such as CERES-MAIZE (Jones and Kiniry, 1986), EPIC (Williams *et al.*, 1984) and CROPSYST (Stockle *et al.*, 1994), however, are able to simulate the effect of water depletion during the growth cycle. They can therefore be effective tools for forecasting the water content of the soil and crop response (Cabelguenne, 1996). Examples of model use for irrigation scheduling can be found in both developed and developing countries. In developing countries, however, such systems have mostly been applied in a commercial context, while in developed countries they have been used by individual farmers. The following paragraphs give some examples of this application of models in developing and developed countries.

McGlinchey (1995) describes a pilot irrigation scheduling project established in northern Zululand on a commercial estate. Meteorological variables were measured with an automatic weather station (AWS), and the data transmitted electronically to the experimental station every week. A model was used to estimate the soil water content on a daily basis. A report on the current soil water status was then generated and advice on when next to irrigate was sent to the irrigator.

Also in South Africa, the PUTU model (de Jager *et al.*, 1983) has been used for irrigation scheduling of many crops. It is used mainly by consultants to provide advice to farmers. Much work has been done on the model to make it user friendly (A. Singels, S.A. Sugar Association, 2000, personal communication). The IRRICANE model (now called CANESIM and partly derived from the CANEGRO model present in the Decision Support System for Agrotechnology Transfer, DSSAT) is also used for irrigation scheduling (Singels *et al.*, 1998) and as a general tool to assist in the agronomic management of sugarcane (Singels *et al.*, 2000).

At a more strategic scale, CROPWAT, a model for estimating crop water requirements, was used by the Food and Agriculture Association (FAO) to develop irrigation guidelines (Penning de Vries, 1990). The CROPWAT approach has been used widely by consultants and others designing new irrigation schemes or introducing new crops that require irrigation. It has recently been modified to use the Penman–Monteith estimate of evapotranspiration, as the original algorithm tended to overestimate crop water use by about 10–15%. Similar irrigation planning approaches have been used elsewhere. For example, in India a crop–soil model was used to prepare irrigation calendars for cabbage, onion, tomato, maize, green gram and mustard (Panigrahi and Behara, 1998).

In Denmark, irrigation scheduling had been based on local experience and rule of thumb, but these proved inadequate to deal with the increased complexity of water management. A PC-based DSS (MARKVAND) was developed in response to a need to find more efficient forms of irrigation in

Denmark due to increasing water demands in different sectors of society (Plauborg and Heidmann, 1996). It is being used to give daily information on the timing, amount and economic net return of irrigation for a wide group of agricultural crops. The model includes conceptual and empirical submodels for crop development, water balance and crop yield. About 100 copies of the system had been sold in 1996 and it was being used by at least 200 farmers.

Similarly in the UK, Irrigation Management Services (IMS) (Hess, 1990, 1996) provided a consultancy service that gave farmers weekly advice on which fields to irrigate, when and how based on the results of computer simulation. IMS employs a simple water balance model with crop evapo-transpiration estimated on the basis of soil, crop and weather factors. It was used between 1984 and 1989 to provide an irrigation scheduling service to farmers and growers in eastern England in conjunction with in-field monitoring of soil water and crop cover (Hess, 1996). When the service began in 1984, it operated on a bureau basis with communication by phone combined with farm visits. Farmers saw this personal touch as beneficial, since it gave them confidence in the service and enabled them to discuss broader issues with the adviser. The idea was that farmers should feel free to call and ask for advice at any time without having to wait for advice sent by mail. By 1990, the use of microcomputers had become more common and farmers began to demand the scheduling packages themselves as this was much cheaper than paying for a consultancy service (Hess, 1996). In time, some farmers eventually felt experienced enough in estimating when to irrigate that they no longer needed to use the model. The IMS programme was also integrated into a whole-farm scheduling system (Upcraft *et al.*, 1989, cited by Hess, 1996) to produce optimum schedules considering the constraints imposed by equipment, labour and water availability. However, this has not yet been developed into a format suitable for use on-farm.

Despite the technological advancements in irrigation scheduling, most irrigators do not use the real-time procedures that have been developed by scientists. However, if irrigation is to undergo an increase in efficiency, made necessary by the increase in water demands from all sectors of society, such tools must be adopted (Tollefson, 1996). The challenge to researchers is to develop economically viable technology that is readily adaptable to producers in rural society. This requires more interactive communication between researchers, extension staff and farmers for improving the transferability and applicability of irrigation scheduling techniques (Tollefson, 1996). In practice, these tools may not be taken seriously by farmers until the cost or availability of water forces them to regard water conservation as a top priority. The other external driver that might achieve the same result, at least in theory, is if a formal irrigation scheduling system is required before an abstraction licence is granted.

9.3 Optimizing Fertilizer Application

Estimating the N requirement of crops has long posed a problem to farmers and their advisers since it varies widely depending on the amount and type of organic matter in the soil, the temperature of the previous season, the amount of rainfall, and the type of crop to be grown. In many countries worldwide, fertilizer recommendations are based on results from experiments carried out over a number of seasons at research stations. We were not able to identify any examples of fertilizer DSSs being applied in developing countries other than in estate agriculture. However, within Europe there are numerous examples that illustrate how these systems might potentially be applied.

In the UK, results from many fertilizer trials have been synthesized by the Ministry of Agriculture, Fisheries and Food (MAFF) and are published as recommendations based on an index system that takes prior land use into account (MAFF, 1983). There are now a number of DSSs that aim to provide more precise advice on the application of inorganic fertilizer and organic manure to farmers in the UK. The most commonly used are FERTIPLAN, MANNER, N-CYCLE, PRECISION PLAN, EMA, SUNDIAL-FRS and WELL-N (Falloon *et al.*, 1999). These fertilizer recommendation systems (FRSs) estimate crop N requirement, either at an average economically optimum yield or at the yield specified by the user. The supply of N to the soil, inputs from soil amendments and losses from the system are then calculated, allowing an estimate of the optimum fertilizer application rate. Only SUNDIAL-FRS and WELL-N are based on dynamic crop models, incorporating a full response to changing environmental conditions (Falloon *et al.*, 1999). G. Dailey (Rothamsted, 2000, personal communication) suggests that SUNDIAL, although as yet used only in the UK, could also be applied to other parts of the world. Currently it is being tested by around 100 UK farmers and consultants, and should be generally available in 2001. Dailey's view is that although the responses of yield to N application rates are sometimes fairly flat, leaching and water stress can occur over relatively short time spans, allowing a dynamic model to provide information that an index approach cannot.

Metherell *et al.* (1997) developed a DSS for the evaluation of P and S fertilizer strategies to assist farm consultants and their clients in determining pastoral agriculture fertilizer policies. Animal production responses to fertilizers were estimated from relationships between soil P and S status, fertilizer inputs, pasture relative yields and stocking rates. The model automatically calculates maintenance and economically optimum fertilizer strategies. The user can also enter his/her own scenarios and evaluate them.

There are no DSSs that cover all crop types or all of the forms of organic waste as well as inorganic fertilizers. Falloon *et al.* (1999) suggested that a fully comprehensive FRS needs to be capable of making recommendations for the application of organic and inorganic fertilizers to a wide range of crops and agricultural systems. However, the needs may vary depending on the user and more restricted DSSs may well be adequate for simple arable farming systems.

For example, in South Africa, SASEX (South African Sugar Association Experiment Station) used a model called KYNO CANE to make fertilizer recommendations for sugarcane with a view to achieving maximum probable yields (Prins *et al.*, 1997). The model has been adapted to quantify the risks of under- and over-fertilizing in monetary terms, an issue which most models previously overlooked. It also aids in the correct choice of fertilizer carrier. It is not clear from the literature as to what extent the recommendations made using this system are taken up by the sugar farmers.

9.4 Multiple Decision Support

Even more sophisticated DSSs have been developed that provide support on more than one aspect of crop management. The GOSSYM/COMAX expert system, operated in the USA, has been cited as a successful application of a cropping model for advising growers on the application of N, irrigation and growth regulators (Boote *et al.*, 1996; Newman *et al.*, 2000). COMAX was designed by the agricultural research service of the USDA and operates by hypothesizing and testing. It hypothesizes a scenario for fertilizing and irrigation and then tests the impact by running GOSSYM with the hypothesized values. COMAX was designed to run at farm level on a PC-type computer. It begins by hypothesizing a management strategy for the grower, but as the season progresses, the hypothesized data for weather variables, irrigation and fertilization are replaced with actual values, with the optimum management strategy being recomputed each day based on the updated information (Whisler *et al.*, 1986).

Pilot testing of GOSSYM/COMAX, as a tool for on-farm management decisions related to N fertilizer applications, irrigation scheduling and timing of harvest aid chemicals, was undertaken with two farmers between 1984 and 1986 (Whisler *et al.*, 1986). Positive feedback from both farmers led to the test being expanded to 19 locations in the mid-south area in 1986. By 1987, there were approximately 70 locations involved across the cotton belt from California to North Carolina with all 14 cotton-growing states participating. In 1988, the pilot test programme had expanded to 150–170 sites. The pilot test enabled the validity and value of GOSSYM as a tool to be evaluated as well as the nature and configuration of a technical support group required to make the system available to growers across the Cotton Belt (McKinion *et al.*, 1989). However, GOSSYM/COMAX, after an initial period of success, met the same fate as SIRATAC and EPIPRE (G. Hoogenboom, University of Georgia, 2000, personal communication), with the number of users of the system in the US corn belt dropping from 400 to 100 over a 15-year period as the farmers learned from the system

(F.D. Whisler, Mississippi State University, 2000, personal communication). However, in 1999, when there were unusual weather conditions with high temperatures and low rainfall after a wet start, some farmers did resort to running old versions of the program. A few farmers have continued using the model even after 15 years.

Newman *et al.* (2000) suggest that the success of GOSSYM/COMAX was probably largely based on the extensive period of farm testing. But as Boote *et al.* (1996) point out, the development of GOSSYM/COMAX was heavily subsidized by the research community. The soybean counterpart, GLYCIM/COMAX, is in more limited use as it has not been released to the public. An attempt is underway to distribute the soybean version to producers (H. Hodges, Mississippi, 2000, personal communication), but will depend on technical support being available to program users.

In Western Australia, TACT DSS was developed in consultation with wheat farmers in the Mediterranean wheat-growing region. The system can be used to support decisions made at the start of the season and provides information about changes in the yield distribution given seasonal conditions to date (Abrecht *et al.*, 1996).

In another example, Wafula (1995) outlines the use of CMKEN, a locally adapted version of CERES-MAIZE, for a variety of applications in the Machakos district of Kenya (see also Section 4.5 in Chapter 4). Simulations, using 32 years of weather data, were used to establish the probabilities of outcomes for combinations of different management variables. Wafula suggested that the model could provide information which can help farmers make choices that are compatible with their socio-economic situation. He believed that it might be successful in helping to improve crop yields for resource-poor farmers where traditional agricultural research had failed. The simulation provided information on optimum sowing dates, varietal selection, and use of N fertilizer. In the case of sowing dates, the model output supported the message that was already being given by extension workers (early cropping reduces the risk of crop failure) but that had, until then, had no quantitative support. The model also demonstrated that the suggested practice of high-density cropping could have a negative effect where there were N limitations (Keating *et al.*, 1993), and thus highlighted the need for moderate fertilizer application. However, many of the resource-poor farmers were unable to apply fertilizer due to the financial cost. Although CMKEN is reported in a very positive light, it is unclear the extent to which the research findings are actually being used in the support of on-farm decision making.

9.5 Deciding whether to Implement Emergency Relief

To assist Albania with an unexpected shortfall in the size of its 1991/92 winter wheat harvest, the US Agency for International Development

(USAID) wanted to determine to what extent wheat imports might be offset with emergency N fertilizer imports. The IBSNAT group was approached by USAID to use their DSSAT models to provide a rapid appraisal of the benefits that could be derived from N fertilizer imports (Bowen and Papajorgji, 1992). Albanian scientists provided weather and soils data which was then used to run the CERES-WHEAT model to test the effect of a single N top-dressing applied at different times during the spring. USAID was then able to use these results to evaluate the potential benefit of imported N and the importance of timing applications. They decided that it was worth it, but unfortunately the time delay in finding an available ship and the transport time meant that the fertilizer arrived in the country too late. Nevertheless, it marked the beginning of a substantial aid package to Albania for improving fertilizer markets and availability in the country (W. Bowen, CIP/IFDC, Lima, 2000, personal communication).

Tools to Support Strategic Decision Making

10

William Stephens and Tabitha Middleton

Institute of Water and Environment, Cranfield University, Silsoe, Bedfordshire MK45 4DT, UK

Strategic decisions relate to the formulation of long-term policies or plans, either at farm, regional or national level. A perceived strength of models is that they can be used to provide information to support decisions especially in times of change when existing experience is no longer applicable or when new opportunities arise. The strength of modelling, in this context, is that impacts over a longer time frame can be simulated taking into account many complex factors in a way that would not be possible using conventional approaches. Models can be used to evaluate the longer-term implications of different agricultural practices or policies. By providing an estimate of the potential impacts of different options, outputs from model-based research can also provide a basis for decision making. In many cases, however, models are described in the literature in terms of their successful application by researchers, but the use of this information to support decision making in practice is seldom mentioned. There are, therefore, some overlaps between the applications reported here and those in the previous chapter on the use of models as tools in research.

10.1 Land-use Planning

Land-use decisions have to be made at all levels of the agriculture sector. Governments must know what policy instruments will bring about particular patterns of land use. For example, what would be the effect of making cheap credit available for agricultural inputs? At the farm level, land-use decisions must be taken to satisfy household criteria and goals such as income security and whether basic nutritional requirements can be satisfied while reducing risk, generating certain levels of cash or maximizing returns to capital investment (Thornton and Jones, 1997). In terms of decision support for land-use planning, agroecological zoning and land evaluation are also important steps in determining the agricultural potential of a region.

Reiterating the point made in Chapter 6, land evaluation is designed '. . . to guide decisions on land use in such a way that the resources of the environment are put to the most beneficial use for man, whilst at the same time conserving those resources for the future' (FAO, 1976). Agricultural problems of over-production in Western Europe and the United States pose quite different problems from those prevailing in many developing countries, where there is a desperate need to match food production with population growth. According to Beinroth *et al.* (1998), long-term sustainable development requires that we find effective ways to assess the potential of land-use patterns and predict impacts and performance under different policy or management options. Moreover, in developing countries, there is a particular need to identify data and tools for efficient, objective and comprehensible regional land-use planning to support multi-party negotiation and consensus building where there are potential natural resource conflicts. These methods should take into account basic information on soils, topography, climate, vegetation, as well as socio-economic variables such as relation to markets, skill of land users, level of social and economic development, etc. This section outlines some of the attempts that have been made to use models in these areas.

Crop models have been used in agroecological zoning in order to provide information on the potential for the introduction of new crops or cropping practices at a regional level. Jones and O'Toole (1987) used the ALMANAC crop growth model to illustrate how such models could be used to meet some of the objectives of agroecological characterization such as matching technology with resources and describing the impact of climate variability on crop yields. Angus (1989) analysed the long-term mean agroenvironment of the Philippines to estimate opportunities for multiple cropping of rain-fed rice using the POLYCROP model. Aggarwal and Penning de Vries (1989) used a simulation model to characterize agroclimatic zones in South-east Asia in terms of production potential for wheat, a non-traditional crop for the region and to identify regions that could be more productive in irrigated and/or rain-fed conditions. Aggarwal (1993) used WTGROWS, a model based on the Dutch MACROS model, and a geographic information system (GIS) to determine productivity of wheat at different locations in India as determined by climate and water availability. Penning de Vries (1990) describes how a model was used to evaluate the suitability of soybean in the Philippines, where it is a new crop. The yield was simulated for rain-fed and irrigated, upland and lowland sites over 20 years, and costs and benefits were analysed to give potential net profit.

In a study aimed at providing information to support policy decisions in Malawi, CERES-MAIZE was used to predict yields in two contrasting locations in the central region of Malawi. A regional analysis using the model linked with spatial databases showed variability in maize yields to be attributable to a combination of soil and weather effects. Nitrate leaching potential at the regional level was also shown (Thornton and Jones, 1997).

In another application, the International Potato Centre (CIP) used the LINTUL (Light INTerception and UTILisation) model for the agroecological characterization of global potato production to help target research at production problems in those regions were potato cultivation is most promising. In 1995, it was used by Penning de Vries *et al.* (1995) in a world food study in which potential and water-limited food production was estimated for the year 2040 for 15 major regions of the world. Similarly, Lal *et al.* (1993) used the BEANGRO model to carry out a regional productivity analysis of beans for three sites in western Puerto Rico. The analysis indicated that optimum cultivar selection, planting date and irrigation strategy varied from one site to another. In this, as in the majority of examples reviewed, there has been little or no attempt to assess the uptake and impact of this information.

In some cases, crop models and GISs have been combined to assess the agricultural potential of a given region or to consider the impact of different options. The use of the Agricultural and Environmental Geographic Information System (AEGIS) for assessing the agricultural potential of land previously used for sugarcane production in Puerto Rico (Beinroth *et al.*, 1998) has already been discussed in Chapter 6. The results showed that tomato double cropped with a cereal would increase profits for land use and would probably decrease risks of erosion and N leaching, compared with a number of alternative cropping systems (Hansen *et al.*, 1998). However, it is unclear as to the extent to which socio-economic factors were considered in this conclusion. Moreover, there is no mention of the involvement of local planners in the work nor of whether it had any impact on agricultural practice.

Stoorvogel (1995) proposes the integration of different models and tools as an effective way to analyse different land-use scenarios and thus inform agricultural policies and economic incentives for sustainable agricultural production. He suggests that the limitations in one model can be compensated for by others, making this an ideal methodology for multidisciplinary research and the integration of socio-economic and agroecological data. The method was tested by analysing different land use scenarios for the Neguev settlement in the tropical lowlands of Costa Rica. Crop growth simulation models and expert systems were used for the description of alternative land-use systems. This was linked to a GIS with land use being optimized using a linear programming model. The simulation was used to look at the effects of: (i) changes in capital availability;

(ii) restrictions on biocide use; and (iii) effect of nutrient depletion on farm income, although Stoorvogel makes the point that an infinite number of scenarios could have been analysed.

10.2 Planning for Climate Change

Recently, due to concern over the potential impacts of the build up of greenhouse gases in the atmosphere, the issue of climate change has moved to the forefront of the global scientific agenda. Simulation models are the only way that the impacts of a variety of potential scenarios can be explored. Such simulations can be used to investigate the effect that climate change will have on agriculture. By considering alternative scenarios, decision makers or planners are in a better position to plan future strategies based on the most likely outcome. This may be extremely important to planners in the developing countries which will be the most vulnerable to the effects of global warming.

Some examples of the use of crop models in this area have already been discussed in Chapter 7. The following cases are some further examples which have all involved the use of the Decision Support System for Agrotechnology Transfer (DSSAT) suite of crop models, which have gained widespread acceptance within the research community. The outputs do provide information that could potentially support decisions on agricultural strategy. However, there is no mention as to whether the results have been used in policy consideration or whether they are merely of scientific interest.

In Bulgaria, climate vulnerability assessments for agronomic systems have been initiated (Alexandrov, 1997). DSSAT version 2.1 was used to predict that an increase of between 5 and 10°C would lead to a decrease in the yield of maize and winter wheat.

- In the Philippines, DSSAT was used in combination with results from four general circulation models (GCMs) to assess the impact of climate change on rice and maize crops (Buan et al., 1996). The results showed both increase and decrease in rice yields according to the variety, while maize yields consistently decreased. This was partly due to increased flooding that would be brought about by an increase of 10% in rainfall. The model was unable to simulate the affects of high winds that would result from typhoons.
- In Java, Indonesia, there is concern that rice self-sufficiency, maintained since 1984, may be threatened by climate change. Three models, including DSSAT, were used to simulate climate change so as to aid policy makers in planning for the effects of recurring droughts and other possible changes. The simulations suggested that changes from 2010 to 2050 could drastically reduce rice yields because of an increased incidence of drought (Amien et al., 1996).

- In Argentina, DSSAT was used to evaluate the potential impact of climate change on the productivity of maize, soybean and wheat, three crops making major contributions to the national economy. According to the results, a generalized increase in soybean and decrease in maize would occur. Regional impacts were varied for wheat, which was likely to increase in the west and east but decrease in the north (Magrin *et al.*, 1997).

- A combination model (CERES-RICE coupled with BLASTSIM) was used in conjunction with weather generators from DSSAT to study the effects of global climate change on rice leaf blast epidemics in five Asian countries. The simulation allowed for analysis of distribution of the disease and estimated yield losses over a 30-year period. The simulated climate change had a significant effect on disease development, although this varied according to the agroecological zone (Luo *et al.*, 1995).

10.3 Crop Forecasting

Crop forecasting operates on a much shorter time scale than that used to predict the effects of climate change on crop production or disease incidence. Large area yield forecasting prior to harvest is of interest to government agencies, commodity firms and producers. Crop forecasting can provide an important tool for agricultural planning in both developed and developing countries. Crop forecasting packages have many potential benefits. They enable policy makers to plan for food security and determine import/export plans and prices, and they enable farmers to plan their marketing strategies and provide a basis on which to make decisions about crop management practices such as fertilizer top dressing, irrigation and fertilizer application (Horie *et al.*, 1992). The decision-making capacity of farmers and resource planners would be improved if they had some means of quantifying risk associated with particular strategies (Bannayan and Crout, 1999). So far, the most widely used methods for operational yield forecasting are based on empirical, statistical or sampling techniques. An evaluation of the relative advantages and disadvantages of different systems however, shows that a combination of remote sensing and crop modelling may provide the most effective method (Horie *et al.*, 1992).

Some such packages are still being developed but appear to have potential. For example, Bannayan and Crout (1999) used the SUCROS model to experiment with real-time yield forecasting of winter wheat in the UK. The results showed that the model was able to forecast final biomass and grain yield with less than 10% error.

In developing countries, there is potential for developing early-warning systems using crop forecasting methodology. Early warning of a poor harvest in highly variable environments can allow policy makers the time they need to take appropriate action to ensure food security in vulnerable areas. Thornton *et al.* (1997) describe how the CERES-MILLET model was used in conjunction with a GIS and remote sensing to estimate millet production in contrasting seasons for 30 provinces of Burkina Faso. They found that provincial yields simulated halfway through the growing season were generally within 15% of their final values. They considered the methodology to have considerable potential for providing timely estimates of regional production of the major food crops in sub-Saharan Africa. However, as far as we know, no operational early warning system has so far been produced.

A methodology was developed using the CERES-MAIZE model to assess drought impacts on maize at an early stage in the season. The index was intended to provide an objective measure that policy makers could use to declare areas as drought-stricken and then implement subsidy schemes on a fair basis (du Pisani, 1987). Although the results proved positive, there is no mention of implementation.

An attempt was made by Abawi (1993) to use long-range weather forecasts based on the Southern Oscillation Index (SOI) linked to a crop harvesting model to predict long-term risks associated with earlier harvesting of wheat in Australia. Since long-range weather forecasts are generally not reliable, Horie (1992) suggests that the ideal crop forecast system should accurately assess current crop status and predict future status and yield under the 'most probable' weather. The most probable weather may be 'normal' climate in a given location with allowances for some deviations. An example of such a model is the SIMRIW dynamic crop-weather model combined with a weather information system. Horie (1992) used this model in Japan and predicted that such models would have an increasingly important role in regional rice yield forecasting in that country. However, successful implementation of the model relied upon the fact that, in Japan, meteorological data such as air temperature, sunshine hours, precipitation and wind-speed had been recorded in 860 sites since 1974. Few, if any, such intensive networks of meteorological stations are available in developing countries.

In Europe, many attempts have been used to use models to forecast yields. For example, the Monitoring Agriculture with Remote Sensing (MARS) project involved the use of crop models for long-term yield predictions in Europe. WOFOST, a general purpose crop simulation model (van Diepen *et al.*, 1988), was integrated with a GIS to produce a crop growth monitoring system for operational yield forecasting in the European Union (EU) (Bouman *et al.*, 1996). However, according to John Taylor (Cranfield University, Silsoe, 2000, personal communication), remote sensing techniques and simpler regression models provide an alternative methodology that can be equally as effective as the use of agrometeorology models. In the 1990s, WOFOST was also used as part of the policy study 'Ground for Choices' to explore regional yield potentials in the EU under different management intensities (NSCGP, 1992). The process involved optimizing land use and production systems under four contrasting economic scenarios, and is discussed in more detail in Chapter 6.

Seligman (1990) states however, that crop models have had poor results as yield predictors, and, despite much research, the more advanced versions of such models are not yet being implemented for commercial use.

10.4 Irrigation Planning

Water for agriculture is coming under increasing pressure from alternative industrial and domestic end-users, so an understanding of the likely future demand for water is needed to develop strategies for water management. Planning for long-term irrigation needs is of great importance in areas of water shortage, where water supply is a potential cause of conflict and trade-offs must be considered between different potential uses and users.

Knox *et al.* (1997) calculated the annual irrigation needs for England and Wales using the Irrigation Water Requirements (IWR, Hess, 1996) model, to help the UK Environment Agency in their long-term water management strategies. The IWR model estimates the daily soil water balance for a selected crop and soil type. For each year of the available weather records, the model outputs data on the crop water use, any irrigation applied and the proportional yield loss due to any water stress. The same approach could be applied in developing countries provided sufficient data are available (J. Knox, Cranfield University, Silsoe, 2000, personal communication). Similarly, Hook (1994) used the CERES-MAIZE, SOYGRO and CROPGRO models, in combination with water use models, to predict yield and irrigation demand in Georgia (USA) for drought years using data from the 15 driest years on record. The demand was assessed relative to the mix of crops grown in the region. This has potential in regions where water resources are limited, making it important to plan the permitted area of irrigated crops and water demand for years of drought. It can be used for strategic planning for irrigation water withdrawals at a regional or watershed level. The approach has since been developed further to provide a practical tool for Environment Agency Abstraction Licensing Officers to be able to validate licence applications from farmers. This has been taken up nationally with operational training support from Cranfield University (J. Knox, 2000, personal communication).

Models have also been designed to deal with the question of irrigation management at a farm level, helping farmers and their advisors to link the strategic and tactical aspects of on-farm water management. For example, as already mentioned in Chapter 4, MacRobert and Savage (1998) describe the development in Zimbabwe of an interactive version of CERES-WHEAT, which searches for the optimum intra-seasonal irrigation regime to maximize the total gross margin for a particular soil, cultural and weather scenario, within the constraints of land and water availability. This optimum regime can then be used by the farmer to plan irrigation management strategies. They illustrate the use of the package by evaluating deficit irrigation techniques which aim to maximize the gross margins (per unit of water rather than land) in large-scale farms in Zimbabwe, where land availability exceeds irrigation water resources.

Passioura (1996) gives an example of the strategic use of the OZCOT cotton model which has been calibrated for a given restricted irrigation area, and can be used to decide what area of cotton should be grown in dry years when irrigation water is restricted. For example, if only half of the water supply is available, the model can give advice on whether it is better to grow, say, only half the normal area of cotton at normal irrigation, or to grow the normal area at half the level of irrigation (Dudley and Hearn, 1993).

LORA is a decision support system developed in France to help farmers and their advisers develop a cropping plan for both the irrigated and non-irrigated areas of their farm (Deunier *et al.*, 1996). The model considers which crops should be irrigated and what their water requirements are. The objective is to obtain the best economic return, taking into account the risks due to climate variability and uncertainty regarding production prices. The model produces an irrigation schedule for each crop and a management choice for each climate scenario. The user can also test his/her own irrigated and non-irrigated cropping plan which the model then evaluates over all climate scenarios and seeks optimal management methods. LORA has been available in France since 1990, and has been modified to take into account changes in the European Union's Common Agricultural Policy (CAP). It is used in three areas: (i) providing direct help in decision making for farmers via their advisers; (ii) studying irrigated systems at a regional level; and (iii) providing training in water management at a farm level for advisers.

10.5 Assessing the Benefit of Proposed New Technologies

Since farmers in developing countries tend to be risk-adverse, they may be unwilling to adopt new, more sustainable, practices when they are uncertain of the immediacy, likelihood or scale of the resulting benefits. Moreover, field data on the impacts of new practices or technologies may give misleading results if the long-term variations of factors, such as weather, are not taken into account. The modelling of proposed changes to the system can provide a longer term view of the potential effects of new practices, and can enable the user to see the probability of the outputs from field experiments being representative in the long term.

Soil and water conservation practices provide a typical example of proposed measures for increased sustainability that are potentially risky for farmers, since they often require high levels of investment in the short term, but provide uncertain benefits in the long term. The examples below, some of which have already been discussed in Chapter 5, show how models can be used to help farmers justify the adoption of new or improved technologies by enabling them to identify the probability of positive yield responses and how they can help to validate the results of field research. It is however, interesting to note that there is little mention as to whether the outputs of this research have been shared either with farmers or extension workers.

Stephens and Hess (1999) used the PARCH model to assess the long-term benefits of runoff control or water-harvesting techniques on maize yield in the Machakos district of Kenya. Using the model with weather data for 30 years enabled the relative effect of different levels of water conservation to be evaluated in terms of their likelihood of success. The results highlighted the pitfalls of traditional experimental approaches in areas with highly variable rainfall. In another study, Freebairn *et al.* (1991) used the PERFECT model to compare, on one hand, a farming system using contour banks, stubble mulching and storage of runoff water for later use, with a second system using bare fallow and no conservation structures in terms of the effect on mean annual runoff, sediment loss and wheat yield. In this study, runoff and sediment loss were greatly reduced, and wheat yield was increased by 14%. Similarly, the Erosion Productivity Impact Calculator (EPIC) model has been used to evaluate erosion consequences of cropping practices and tillage (Williams *et al.*, 1984).

Modelling can also help decision making on the optimum design of soil water conservation practices. For example, Stroosnijder and Kiepe (1997) used the DUET model to evaluate various types and frequencies of conservation tillage, aimed at better infiltration of water into the soil, on the growth of millet in the Sahel. They also used a model called SHIELD to determine optimum row spacing in contour hedgerows. However, it is not reported if the results from this work were disseminated to the farmers.

10.6 Planning Optimum Farm Management Strategies in Collaboration with Extension Services and Farmers

In the last 10 years, there has been a realization that models may have a useful role to play in planning strategic farm management with extension workers and farmers. This approach is relatively new, and is still being actively developed (Seligman, 1990). However there are already a few examples of its application both in developing countries as part of internationally funded development projects and in a more commercial context in developed countries.

   Two examples come from Australia. As early as 1987, Kingwell and
Pannell (cited in Seligman, 1990) describe how the MIDAS model, which
is based on simple biological relationships, was being used experimentally
for planning optimum farm management strategies in collaboration with
extension services. FARMSCAPE (Farmers, Advisers and Researchers
Monitoring Simulation, Communication and Performance Evaluation) provides a more recent example of the experimental use of a combination of
hard and soft systems methodologies to support farmers (McCown et al.,
1998, cited in Newman et al., 2000). The FARMSCAPE approach, used for
a dry-land crop production management system, was an alternative to traditional DSSs since it combined simulation (hard system) with participatory interactions with the farmers and advisers. It was hoped that the farmer
would gain the benefit of recommendations and also learn in the process.
The simulator provided the opportunity to compare options for the forthcoming season and was used by the farmer, researcher and adviser as part
of a discussion focused on farm planning. An evaluation of the process
showed that, when taken in context, the simulation helped participants to
gain insight into the way their production system functioned, and increased
their experience in tactical decision making (Newman et al., 2000).
   In a developing country context, an example comes from the Hindu Kush
Himalayas, in which ten Asian countries participated in a project which
used field scale modelling for decision support in participatory watershed
management. The project involved training and workshops, and was so
well received by the participants that China translated the manuals into
Chinese and set up its own training course (Rajan Muttiah, Texas, 2000,
personal communication). A similar process was described by Beinroth
(1998), already discussed in Chapter 6, of using AEGIS to assess the feasibility of small irrigation projects for watersheds in the Colombian Andes
as part of an interactive irrigation planning exercise. The model could be
used to explore trade-offs between domestic water requirements, irrigation
demand and downstream use, and be able to support the process of consensus building during community discussions by enabling new positions
to be simulated and evaluated in an iterative manner.
   In Zimbabwe, the Agricultural Production Systems Simulator (APSIM)
model is currently being applied in an Australian-funded 'risk management'
project to help farmers, policy makers, extension agents and researchers
improve their understanding of the trade-offs between different crop and
cropland management strategies under scenarios of climatic risk (P. Grace,
CIMMYT, Mexico, 2000, personal communication). APSIM was chosen
because of its ability to deal with complex interactions between climate,
soil fertility, and crop and residue management. Rather than focusing on
an optimal strategy, the scenario analysis focused on practices that would
be feasible and productive in a context where farmers are managing
multiple fields under tight labour and capital resource constraints. The
modelling aspect was important since it would not have been possible to

undertake such analysis either on-farm or at a research station in a reasonable time frame. The project was designed to open up new perspectives in a training setting. It demonstrated how simulation models can help in thinking about the options open to resource-constrained small-holders combined with dialogue with farmers and extension workers. The participants' awareness of model limitations was helpful since they were aware of the need to 'reality check' the outputs of the models, and to consider in particular the implications of constraints not captured in the models. Although there were deficiencies in terms of modelling household constraints, the study was able to highlight the challenges faced by small-holders when investing in fertilizer inputs and it provided pointers for future on-farm research directions.

Why has the Uptake of Decision Support Systems been so Poor?

11

William Stephens and Tabitha Middleton

Institute of Water and Environment, Cranfield University, Silsoe, Bedfordshire MK45 4DT, UK

Although the preceding chapter provides a range of examples demonstrating the potential applications of crop models for decision support, there is little concrete evidence of their practical application beyond the research domain. Ideally, one would be able to use the lessons from past successes and failures of decision support system (DSS) applications in order to define a realistic and practical future role for models in this capacity. However, the limited documentation on the subject makes it difficult to draw clear lessons from the cases reviewed.

The problem is not a lack of literature on models for decision support *per se*, but rather the nature of this literature. Published work on the subject generally focuses on issues of interest to the research world, such as descriptions of model development and validation. This may be natural, since journal publications tend to be aimed at other researchers in the field, and those who develop models may be primarily interested in whether the models work technically. However, it means that failures tend to be overlooked, while applications beyond the research world go unreported, since commercial enterprises and non-governmental organizations (NGOs) do not often publish their work. As a result, there is very little information available on the practical application of specific models, when compared to the wealth of literature on their development and potential application.

This lack of published evidence of model application may mean either that crop models are not being used for decision support, or that their use is not documented. However, anecdotal evidence of model use by consultants around the world supports the view that many cases of application are not reported since the simulation results belong to the clients. For example, Decision Support Systems for Agrotechnology Transfer (DSSAT) models have been used by modelling consultants as tools to aid decision making in South Africa (Abraham Singels, S.A. Sugar Association, 2000; Dewi Hartkemp, CIMMYT, Mexico, 2000; Andre du Toit, South Africa, 2000; personal communications). The cases cited below all come from South Africa's Grain Crops Institute which has used the DSSAT software package for decision making on a variety of projects. It is important to note that the modelling work was carried out on behalf of the clients by the modelling consultancy. The clients did not use the models themselves.

- DSSAT was used to simulate the production potential and risk of maize on two farms for an organization wishing to buy a commercial farm for small-scale farming development. The modelling helped them in deciding which farm to buy.
- DSSAT was used for a fertilizer company to gauge the optimum level of N application for a particular field.
- CERES-MAIZE was used in yield estimation for the Orange Free State Department of Agriculture. This data is one of the inputs of NCEC, the outputs of which are used in FAO and SADAC early-warning systems.
- A simulation study of the impact of climate change on South African maize production. This will be used as a basis by a mitigation team for planning to minimize the impact of climate change.
- DSSAT was used to determine the maize potential and risk of rehabilitated soils for a mining company who have bought land from farmers with the understanding that at the end of the open mine activities the land will return to farming activities. It will also help the mining company monitor whether they are on target in restoring the original potential.

Although DSS is clearly being applied, the success of these systems is still under debate (Abraham Singels, S.A. Sugar Association, 2000, personal communication). It is difficult to ascertain whether these accounts represent a large quantity of undocumented applications, or whether they are merely exceptions.

Despite the lack of detailed documentation on the specific application of crop models, there is no shortage of review papers debating whether or not they are useful outside the research domain (e.g. Passioura, 1973; Seligman, 1990; Hamilton *et al.*, 1991; Philip, 1991; Bouman *et al.*, 1996; Cox, 1996; Passioura, 1996; Sinclair and Seligman, 1996; Newman *et al.*, 2000). Their judgement on current applications is not positive. For example, both Seligman (1990) and Bouman *et al.* (1996) consider that the performance of crop models in farm management and decision support has been a particular disappointment, since there are surprisingly few examples of successful applications despite their wide use in research studies, and the fact that many of them have been specially tailored for use by farmers or extension personnel. Interestingly, this problem is not unique to the agricultural sector. For example, Seligman, (1990) quotes Shortliffe *et al.* (1979):

'A recurring observation as one reviews the literature of computer-based medical decision making is that essentially none of the systems has been effectively utilised outside of a research environment, even when its performance has shown to be excellent'.

Unfortunately, more than 20 years later, our own literature review reinforces their finding that the practical application of crop models for decision support has been poor in relation to the number of models developed and their potential as claimed by their authors. The few documented examples of practical applications come either from developed countries, or from the commercial sector of developing countries. To date, the highest recorded uptake has been of models aimed at improving on-farm tactical decision support such as disease and pest control (e.g. SIRATAC and EPIPRE), and irrigation scheduling and water management. Even here the examples are limited. In terms of strategic decisions, there has been much investment in the use of models to predict impact of climate change or to investigate more sustainable land management practices, however, it is unclear as to the extent to which the outputs from this work have actually been used to support planning decisions. Attempts to use models for the direct benefit of small-scale farmers and extension workers in less-developed countries (LDCs) appear to be still in their infancy.

Problems with model implementation experienced in developed countries are likely to be more prevalent in the developing world, where there is poor access to many of the resources needed to maintain such systems (e.g. computers, data, well-trained staff, etc.). Even in Australia, where much of the modelling research and development is based, adoption has been low, with only 5% of Australian farmers in 1990 using DSSs (Hamilton *et al.*, 1991). Bearing this in mind, it is unsurprising that application in LDCs is poor. This may be due to the fact that computer-based decision making is such a new tool in developing countries that it is hardly used by policy makers, let alone farmers (Rajan Muttiah, Texas, 2000, personal communication). However, Struif Bontkes (IFDC–Africa, Togo, 2000, personal communication) claims that even 20 years of effort to introduce agricultural models as a tool to aid decision making in West Africa has not resulted in significant adoption. Nevertheless, as computers become cheaper and more readily available, this situation may change – a recent report by the US National Agricultural Statistics Service, for example, found that nearly 55% of farms in the US had access to a computer in 2001 compared to the 1999 level of 47% (NASS, 2001).

Technology adoption is clearly a complex issue. There is often a wide range of constraints to adoption, any one of which can prevent the uptake of models for decision support. These may interact or reinforce each other so that the removal of one without considering the others will not necessarily guarantee model application. These constraints occur at a variety of levels and may be technical, intellectual, or operational in nature. In the following sections, we discuss the major reasons for non-adoption of DSSs which have been mentioned in the literature and list-server survey, some of which are highlighted in Box 11.1. A fuller discussion of model limitations in general is in Chapter 14.

Box 11.1. Reasons for poor adoption of decision support system (DSS) models. Sources: Cox (1996), Newman *et al.* (2000).

- Unclear definition of clients/end-users
- No end-user input prior to or during the development of DSS
- DSS does not solve the problems that client is experiencing
- DSS does not match their decision-making style
- Producers do not trust the output due to lack of understanding of the underlying theories of the models utilized
- Producers see no reason to change current management practices
- DSS does not provide benefit over current decision-making system
- Limited computer ownership amongst producers
- Lack of field testing
- Cannot access the necessary data inputs
- Lack of technical support
- Lack of training

11.1 Model Construction Constraints

One of the main constraints to the use of DSSs is that the available models are not perceived to be applicable to the problems experienced by the decision makers. In their evaluation of the PARCH project, Stephens and Hess (1996) noted that in some cases the target clients could see no relevant application for the model in their work. This problem is not unique to PARCH. The reasons for such a perception are varied. In some cases, it may be based on poor understanding of the relevance of models as DSS tools. Such a problem can be overcome with a good training course. However, in other cases, low uptake may be based on more fundamental issues such as the fact that the model does not answer the questions that interest the client.

11.1.1 Inappropriate focus on scientific issues

The problem of inappropriate focus is rooted in the model development process. It is widely accepted that a major reason for poor model adoption in DSSs is linked to the undue emphasis of many models on problems of a scientific rather than a practical nature, resulting in failure to address the problems that the decision makers are facing (Hilhorst and Manders, 1995;

Cox, 1996; Knight, 1997; Struif Bonkes, IFDC–Africa, Togo, 2000, personal communication). For example, many models designed initially for research purposes aim at accuracy in their representation of the system, while farmers may be more interested in other factors such as satisfying a limited time constraint or using only the minimal information required to reach decisions (Will Coventry, University of Queensland, 2000, personal communication).

An inappropriate focus on model research and development, rather than on finding the best ways to solve those problems of interest and concern to the client, means that the whole model development process is starting on the wrong foot (Cox, 1996). Instead of focusing on the needs of the client, and whether DSS would be the most effective intervention (for example, by asking how it compares with other ways of doing the same thing), time and money are invested in finding applications for models that were not designed with a specific client clearly defined. Indeed, market studies are often only carried out after the model has been developed when there is concern with poor adoption (Cox, 1996). This mismatch between the scientific questions asked and the practical needs of the end-users clearly reduces the chances of the model being practically useful. However, it is not an insurmountable problem if those developing the models are willing to learn from their mistakes. Doyle (1997) considers the issue in the case of pest control (Box 11.2).

11.1.2 Not relevant to routine on farm decisions

Cox (1996) believes that the expectation that farmers could usefully exploit DSS technology is seriously flawed because they rarely provide information on improving performance in routine situations rather than the exception. Farm management is mostly routine and major changes are seldom contemplated. If only major crises such as price fluctuations or drought are likely to prompt change, then the farmer may not consider resorting to technological support. Despite this view, there is, however, evidence of the use of models having a profound impact on yields in the absence of any major crisis. Sugarcane research in South Africa has shown huge responses to supplementary irrigation. Irrigation scheduling has also been seen to have a drastic effect on crop yield. However, while commercial interest in irrigation scheduling and crop yield prediction is high, it is unclear how the benefits from this technology can be shared by the 47,000 small-scale farmers in the region (A. Singels, S.A. Sugar Association, 2000, personal communication).

The issue of routine decisions is clearly less relevant to other groups, such as local government staff and national policy makers, who may need to consider strategic issues of land-use planning, or changes in agricultural practice in the face of climate change, or other environmental influences.

Box 11.2. The mismatch between model enquiry and end-user needs (from Doyle, 1997).

Although weed models have provided a framework for developing pest and weed control strategies, models aimed at Integrated Crop Protection (with a few exceptions) have tended to direct their attention to the scientific question of what, rather than the practical question of how. Weed management models have primarily addressed three questions:

- What is the relationship between the level of weed infestation and crop losses?
- What level of control is required to contain the infestation or totally eradicate the weed?
- What is the level of weed or pest infestation above which control measures are justified?

However, those interested in the practical implementation of Integrated Crop Protection need answers to a different set of questions:

- How is it possible to promote the more selective use of herbicide, while ensuring economically acceptable levels of weed control?
- How is it possible to minimize the environmental impacts of herbicides through the use of biological and physical control techniques?
- How can the economic risks to growers of switching to non-chemical controls be minimized?

This 'mismatch' between the aspirations of the integrated crop protection programmes and the main areas of model enquiry may mean that many of the outputs are not integrated into the farming sector.

11.1.3 No clear advantages

Farmers need to see that computer systems have obvious advantages over other decision-making methods which they have been using for years. In order to be of interest, models must solve problems not readily solved by current rules of thumb (Hamilton *et al.*, 1991). Clearly, farmers do not need computer-based DSSs to make decisions. They may help to make 'better' decisions, but farm decisions can be and are made without them. Moreover, there is little point using models to emulate farmer decision-making styles, since the farmers already do this adequately. The strength of models is that they are able to deal with complex interactions and produce precise results (W. Coventry, University of Queensland, 2000, personal communication). However, Cox (1996) suspects that the process models on which DSSs are based have inadequate resolution to distinguish much of what interests practitioners. Anderson and Hardaker (1992), on the other hand, suggest that model focus on accuracy is misplaced since in farm planning, making an optimum decision does not greatly increase benefits and the costs of making a suboptimal decision are usually slight. This is not to suggest that inaccurate models are desirable, but merely that there is little point pursuing the absolute optimum – as long as one is reasonably close to the optimal region – since there are few benefits from refined optimization.

11.1.4 Questions are asked in the wrong way

Not all commentators share the view that models ask the wrong questions. The issue may be more related to the decision-making style of the producers (Newman *et al.*, 2000). Dewi Hartkemp (CIMMYT, Mexico, 2000, personal communication) believes that existing models *can* usefully address the problems that decision makers face, and that the problem is one of language rather than content. He suggests that since models were initially designed by scientists to explore possible innovations, 'they do not talk the way that decision-makers do'. The fact that many of them still look like research tools may be off-putting. However, he does not see this as an insurmountable problem since a base model can be altered to fit the needs of the client. He cites the example of CROPGRO soybean and PCYIELD soybean (Welch *et al.*, 1999), which have entirely different interfaces but contain the same underlying model.

11.1.5 Failure to define target user group or client

If the system is not answering the right questions or not answering them in the correct manner, then it is unlikely to be used. However, it is not possible to be sure that the right questions are being asked if the needs of the client are not known. Newman *et al.* (2000) suggest that because so many DSSs have been designed to answer scientific questions, the beneficiary of the system is often unclear at the development stage. Failure to clearly define the end-user can result in a system that does not meet the needs of any group. Definition of the client is vital if the system is to be appropriate and usable. Different strategies will be required depending on the social and technical development of the client group (Itier, 1996). It is not enough to focus on the problem alone, since a system that will work for highly mechanized agriculture in the West may be quite unusable for farmers in developing countries. For example, the success of SIRATAC and EPIPRE could not be repeated with small-scale LDC farmers as they would have difficulty gaining access to a phone to communicate their results to the consultants. Clearly, the needs of farmers will vary between developed and developing countries, as will those of policy makers and planners at a national and regional level. This means that universal DSSs are inappropriate. For example, many DSSs are aimed at output maximization in high-input farming systems whereas farmers in LDCs may be considering risk minimization in low-input systems. Different systems are required to confront the different types of problem faced.

11.1.6 Failure to involve the client in the model development

Defining the client is not the only important step to be taken in order to increase the chances of successful model application. If there is poor communication between the client and model developers then their needs may still be overlooked or misunderstood. Indeed, Tollefson (1996) found that the greatest obstacle to the adoption of irrigation scheduling (or other simulation systems) was the lack of interactive communication between researchers, extensionists and farmers – a view echoed strongly by McCown (2001). Researchers in the past have developed information, which has then been passed through various extension mechanisms to the producer, but there has been no mechanism for producers to feed back their opinions. Producer input is vital if the work is to be relevant and utilized. Participatory work integrating the ideas of researchers, extensionists and farmers may be the way forward in DSS development and application since it ensures that the focus of the model remains on relevant issues.

11.2 Marketing and Support Constraints

A recent study carried out on the performance of information technology (IT, including DSSs) in the UK, drawing on 900 man-years of professional work, found that 80–90% of IT investments did not meet their performance objectives and 40% failed or were abandoned. Up to 90% of the problems were considered to be organizational rather than technical in nature (Clegg *et al.*, 1996). This is supported by other research, both in the developed and developing world. Researchers from the University of Western Sydney, who carried out an in-depth participatory analysis of the limited adoption of SIRATAC (Macadam *et al.*, 1990), suggested that reasons for non-adoption were symptoms of more fundamental organizational problems of policy, direction, management, marketing methods and funding. In other words, many of the problems did not lie with the DSS itself, but with the way that it was presented and marketed.

11.2.1 Poor marketing

The marketing system is of key importance in the success of DSSs (Hamilton *et al.*, 1991). Even if a DSS answers appropriate questions, the marketing strategy and price will affect the client's willingness to apply such a system. Appropriate marketing is important if the model is to be widely applied.

Many systems have been developed within the academic community where there is little or no experience of marketing. This can lead to poor marketing or distribution strategies and inappropriate pricing. For example, if the system is free, then it may be perceived as having no benefit and therefore be rejected. If it is too costly, then farmers may not feel that it is worth paying for if they are already making adequate decisions, or that they can get cheaper adequate advice elsewhere (Knight, 1997). If the use of a new system requires a high investment (buying hardware and software), poorly resourced decision makers may well feel unwilling to take the risk unless they are certain that the benefits are worthwhile. A trial period may help to solve this problem.

Since there is often only a niche market for DSS products, commercial software companies are unlikely to be interested. Issues such as upgrade and maintenance need to be considered in such a fast-evolving field. Models should not be sold in the same way as a book or other objects as this ignores these issues (Hamilton *et al.*, 1991). Rather they should be sold as a whole package, including education/training as well as technical and intellectual support.

The importance of marketing strategy was noted by the developers of LORA (a water scheduling system) when a change in marketing strategy led to a change in uptake (Deunier *et al.*, 1996). The initial marketing policy proved unsuccessful. The model was marketed via a computer program publisher, but the limited market for the DSS and the difficulties the publisher had in training potential users led to failure. LORA is now distributed to potential users by way of contract, including training on its design principles and operation.

The marketing effort used to promote PC Plant Protection in Denmark highlights the possibilities for success with 1400 licences sold in 1995 following a price cut from $350 to $100 and an advertising campaign in local and national agricultural newspapers (Murali *et al.*, 1999). However, the subsequent rate of uptake of less than 200 licences a year despite a targeted mail shot to 22,000 farmers suggests that increasing market resistance rapidly ensued and that initial success is no guarantee that this will continue as the 'early adopters' amongst the farming community may have already all responded.

11.2.2 Poor dissemination

In developing countries, the issue of dissemination is also of great importance. DSS for sustainable land resource management may not be marketable in this context. Where resources are limited, institutions may be unwilling to invest in a system unless they are sure that it will be of benefit. Moreover, if a large proportion of the target group is unaware of a product, they are unlikely to search for it. Increasingly, models are being placed on the Web and can be accessed freely, but this immediately excludes those unfamiliar with the Internet, as well as those who do not know where to look.

Dissemination must involve more than just promoting the package and then leaving. For example, the PARCH model had been disseminated to the appropriate research institutes in the sense that they had been given the software and offered a short training course. However, one of the reasons for its poor uptake was the fact that participants did not know where they could go for technical support (Stephens and Hess, 1996). The obvious response to this would be to include technical support in the dissemination programme. Unfortunately, previous attempts to do this by providing the services of an expatriate scientist were unsuccessful; when the expatriate scientists withdrew, modelling activity rapidly decreased (Stephens and Hess, 1996).

It may not always be the model itself that requires dissemination. In cases where scientists carried out the simulation, the information produced needs to be disseminated. If such outputs are to be practically applied they must be shared with those who can use them, not only with other scientists. For example, few papers describing models that have been used for analysing impact of climate change or other issues of strategic relevance mention whether the information was shared with potential decision makers.

Hamilton *et al.* (1991) use a product life cycle graph to show investment in model development and dissemination (Fig. 11.1). In many cases, it would appear that the models remain at the development stage, since, after

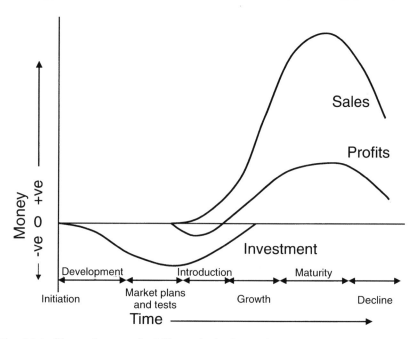

Fig. 11.1. Stages in a product life cycle (redrawn from Luck and Ferrell (1979) in Hamilton *et al.* (1991)).

validation, the 'development project' is seen to have been successfully completed. Since no further investment is made in the dissemination, the model remains on the shelf. If the cost of the dissemination process is not clearly understood and incorporated into the whole project, then it is unsurprising that the model is not used. The report by Clegg *et al.* (1996) suggests that substantial resources need to be invested in the human and organizational factors of IT dissemination and that these may amount to 50% of the total cost of development.

11.2.3 Lack of training in the development and use of software

In many cases the training provided with new software is inadequate or non-existent. Software may initially daunt new users. Lack of training in the development and use of DSSs will restrict the ability of many practitioners to benefit from the potential support that they can offer. Understanding and confidence can only be gained through experience and training in DSS use, to allow farmers and other decision makers to discern what it can and cannot achieve, and the extent to which outputs are valid. Not only do many people in developing countries lack familiarity with models and the modelling process, they often have no experience even of using computers. Because of this, the cost of using a model is often too high, even for institutions, because learning to use the model takes time and requires familiarity with computers. For example, Mathew Laurenson (NARC, Japan, 2000, personal communication) suggests that it takes at least 1 month for a new user to become proficient with Windows-based applications. This is borne out by our experience with students at Cranfield University who have not used computers before.

In another example, Bernet *et al.* (2001) describe the development of a farm household DSS to tailor extension to different production contexts. The model covers three ecological zones in the Peruvian Andes and aims to identify appropriate strategies for raising incomes, especially through milk production. The model outputs suggest that different interventions are required in the three ecological zones to reflect the varied production systems. These include sprinkler irrigation to increase fodder production and erosion protection measures, especially on sloping land. The utility of the model must, however, be questioned, as the authors admit that users 'must be competent in using Excel features and linear programming' and further that 'students who used the model for their BSc theses were not well enough prepared to run the model themselves'. This candid admission suggests that, despite the authors' aspirations, the model may well remain a research support tool rather than contributing to farmers' decision-making processes.

11.2.4 Short shelf-life of DSS software

Software has a short shelf-life, and unless new features are added to the products and they keep pace with advancing technology, they will become obsolete. There are two aspects to this – a DSS may become obsolete due to changes in the computer operating system needed to run it, or it may become obsolete due to there no longer being a need for it, perhaps due to its own success.

The rapid evolution of computer operating systems has meant that programs must be constantly updated. Programs that were originally developed for MS-DOS computers are increasingly unlikely to run on today's computers using Windows, and there is no guarantee that Windows will not eventually be succeeded by a superior operating system, necessitating further changes and updates. Keeping programs updated requires a continuing investment of time and money. Often this is not a trivial matter – the move from MS-DOS to Windows involved a change in programming paradigm from sequential programming in which a program had a start and a finish with a series of instructions in between, to event-based programming in which subprocesses of a program were initiated by input from the screen or mouse of the computer.

Similarly, decision support packages may also become obsolete when farmers using them develop rules of thumb and are able to make decisions without them. This was demonstrated clearly in the EPIPRE and SIRATAC examples (Hamilton et al., 1991). Having such a short shelf-life adds considerably to the cost of developing such packages. These systems are unlikely to be profitable since they require a substantial research effort and can take a long time to develop before they reach the market (F.D. Whisler, Mississippi State University, 2000, personal communication). In terms of pest and weed management, the problem of short shelf-life has been compounded by the fact that many pest, weed and crop disease problems are developing very rapidly and by the time the models have been developed and potential solutions generated, the solutions have ceased to be relevant (Doyle, 1997).

However, rules of thumb, which are usually based on average years, are also rendered invalid when conditions change (see the example of GOSSYM/COMAX), and DSSs may be useful in such situations to adapt or modify these.

11.2.5 Institutional resistance

Even when models themselves are well designed, there may be political or institutional constraints, which mean that the improved DSS is not used or that the outputs are ignored. Without an organization able to respond to the output of the model, any implementation is likely to result in failure.

Through analysing the implementation of the INCA irrigation scheduling software on schemes in Sri Lanka, Thailand, Bangladesh, the Philippines, Jamaica, China and Turkey, Makin and Cornish (1996) realized that although DSS software is designed to improve the way that managers make choices between alternatives, this is often anathema to bureaucratic and administratively oriented agencies where staff are more accustomed to following a single course of action.

11.3 Technical and Operational Constraints

Technical and operational constraints relate to the actual implementation of the model and its running on a routine basis. In developing countries, technical issues are likely to play a more important role in hindering technology uptake than in developed countries, due to the more limited support available. The main technical constraints are outlined below.

11.3.1 Poor access to hardware and software

Access to software and hardware is vital if models are to be used. This may be limited in poorly funded organizations and is certainly not an option for small-scale farmers. However, Knight (1997) suggests that the potential for implementation of DSSs is likely to increase in the future as computers become cheaper, more powerful and more widely used. Evaluation of the uptake of the PARCH model in southern Africa (Stephens and Hess, 1996) showed that, apart from the situations where disks were not accessible, few technical constraints prevented the uptake of PARCH. Lack of suitable computer hardware was only rarely a problem and most organizations were well equipped with computers.

11.3.2 Lack of data

In developing countries, inputs required for model use are often not readily available (i.e. in digital form). This lack of readily available data at all levels (local, national, regional) is a strong limiting factor (Bouman *et al.*, 1996; Rajan Muttiah, Texas, 2000, personal communication; Mathew Laurenson, NARC, Japan, 2000, personal communication). For example, the quantity and quality of data in West Africa is so low that it is difficult in most countries even to obtain reliable production data necessary for input into famine early-warning systems. As a result, remote sensing has had to be used as a substitute (William Payne, Oregon, 2000, personal communication). Reducing data requirements is problematic if models are to produce accurate outcomes. This is particularly true for agricultural and natural resource issues which are characterized by complex interactions (Walker and Lowes, 1997). As long as the necessary data are unavailable or difficult to obtain, model application at all levels will be limited.

11.3.3 Limited validity of models in LDCs

Some reviewers suggest that crop models are not valid for use beyond the specific set of conditions for which they were designed. At their best, their outputs are only relevant to the conditions that were used to calibrate them. According to Bouman et al. (1996), most crop models are developed for strictly defined hypothetical production situations with uniform fields. When they are then used in real field conditions where several limiting factors may occur simultaneously, they fall outside of their domain of validity. Octavio Castelan (University of Edinburgh, 2000, personal communication) suggests that, since the validation process normally stops at field trial sites, and models are seldom tested in the field with farmers, their ability to be applied to development and poverty alleviation is zero, because it is not known how models perform under relevant conditions.

11.3.4 Poor user interface

Surveys of DSSs reported by Greer et al. (1995) suggest that the complexity of the user interface is one of the most limiting factors in their uptake and use. Ludmila Pachepsky (Maryland, 2000, personal communication) believes that if models embedded in user-friendly interfaces were made easily accessible in the public domain, then 90% of the uptake problem would be eliminated. A survey of farmer opinion in the UK by Smith et al. (1997) was carried out to ascertain what farmers want from DSSs for fertilizer application. Farmers requested that default values be available for all model inputs, that inputs should be entered via a Windows-based menu (for clarity) and tabular format (for speed), that users should be able to select which units they wish to use, and that the software be fully supported by context-sensitive help. The system should have a hierarchical structure allowing access to fixed parameters, and be compatible with commonly used recording packages. Recommendations should be provided both for the field and across the whole farm. Simulations should be able to be easily rerun using more recent weather data. Processes underlying recommendations should be illustrated with flow diagrams, bar charts and pie charts.

Many others (Knight, 1997; Newman et al., 2000; Abraham Singels, S.A. Sugar Association, 2000, personal communication) agree on the need to create user-friendly interfaces if models are to become more successful as DSSs, but do not consider this to be the main problem in terms of model application. For example, in their evaluation in southern Africa, Stephens and Hess (1996) found that the ease of use of the PARCH model itself very rarely prevented people from using it. In fact, it generated considerable interest amongst participants and was widely regarded as intuitive and easy to use. However, this may have been related to the ease of running the model rather than the ease of obtaining and entering input data, since not one of the institutions visited was using it.

Despite the importance of technical constraints, there is much evidence to suggest that the main factors inhibiting the use of models relate to generic issues of technology development and dissemination as well as organizational issues, such as marketing, rather than to technical problems (Zadoks, 1981; Macadam *et al.*, 1990; Cox, 1996; Stephens and Hess, 1996).

11.4 User Constraints

Some constraints are due to problems that the users themselves may have in their own perceptions of models.

11.4.1 Lack of faith in models' predictive powers

Even where decision support models ask the right questions, many decision makers have no faith in their predictive powers (Struif Bontkes, IFDC –Africa, Togo, 2000, personal communication). This problem may be particularly acute in developing countries. Hamilton *et al.* (1991) believe that the real decision makers in agriculture (i.e. the farmers) have a healthy suspicion of 'black boxes' which supply a simple answer to complex management decisions. William Payne (Oregon, 2000, personal communication) supports this view, and says that the lack of faith is justified in LDCs since the data available are so limited as to make the model outputs questionable. If farmers do not have confidence in their reliability and some understanding of what is incorporated into DSS packages, these tools are unlikely to gain acceptance. Ludmila Pachepsky (Maryland, 2000, personal communication) believes that such lack of faith only arises where there has been poor training. By showing the clients clearly how the model can be used, the assumptions upon which it is based, and the limitations that it contains, then they will no longer need to base their views on faith in a black box, but rather on an understanding of the tool. One way to overcome the problem is, therefore, to train model users so that they understand the real potential of models to help them in their work. Another approach is for model developers to work much more closely with end users so that 'best practice' models can be developed in the context of practice rather than theory (McCown, 2001).

11.4.2 Technophobia

Although training may, in the long term, solve many problems such as poor faith and limited understanding of what the model can achieve, many model makers and researchers fail to appreciate the fear that new technology can instil in those unfamiliar with it. In developing countries where computer technology is rarely available at the household level, many users may feel daunted by the task of using a computer for complex decision making. It is often hard for those comfortable with such technology to understand such fear and if they are involved in training, they may start on a level that is already beyond the understanding of their clients. However, pride may result in the client pretending to understand, but later preferring to ignore the new technology (i.e. the DSS). Where the client is already busy and is perhaps dubious about the benefits of model use, the investment in time required to overcome technophobia and familiarize themselves with model use may be too high.

11.5 Other Constraints

There are certain factors which are not direct constraints to adoption but may hinder the long-term potential for the application of DSS in developing countries.

11.5.1 Importance of the policy environment

In many cases on-farm use of DSSs in developed countries has been based on the need to improve the effectiveness of agricultural practices to ensure that they are compatible with increasingly stringent environmental legislation. Farmers in Europe are under increasing pressure to reduce the amount of chemicals that they add to their crops (herbicides, pesticides and fertilizers) while maintaining a reasonable yield. There is also increased pressure on water resources. In this context, models that support decisions for integrated pest management (IPM), fertilizer application or irrigation scheduling, will be of great value. This may not be the case of small-holder farmers in developing countries who are not currently under pressure to comply with environmental policies. Such demands may, however, trickle down through donor agencies and develop in time.

11.5.2 Problems with acting on decision support

DSSs tend to be based on the view that decision-making is a purely rational process with demonstrably correct solutions. In reality, natural resource management problems are often inherently political in nature (Walker and Lowes, 1997), and being able to see the options only in terms of the biophysical environment may not be the factor limiting sustainable development. In many cases, small-scale farmers in developing countries may not be able to implement optimum land management practices for external reasons, and therefore DSSs will not help them. However, this does not negate the potential for computer models to support sustainable natural resource management in terms of problem formulation, task analysis, and making effective use of available data and rational use of results of analysis (Walker and Lowes, 1997).

11.5.3 Lack of validity of model outputs

Some model developers and critics believe that the assumptions upon which models are based are still very shaky, and that more understanding of natural and social systems and their complex interactions is needed before models can be useful in decision support. Jones *et al.* (1993), reviewing the DSSAT project, concluded that although the possibility and value of such systems had been demonstrated, DSSs for agriculture was still in an early stage of development because problems faced by decision makers are far more comprehensive than those that can be addressed by the current DSSAT. Bouman *et al.* (1996) suggested that the operational use of deterministic models that can handle the complex systems that typify actual farming conditions is a long way off, since major gaps still exist in knowledge, making it impossible to use mechanistic models directly for farm level applications.

Walker and Lowes (1997) suggest that although DSSs have been usefully applied in engineered, or simplified, intensively managed systems, their use for problems of sustainability at the landscape level are constrained by the fact that these systems are characterized by complex interactions which are only partially understood, and for which there is limited data available. Moreover, Bouman *et al.* (1996) suggests that the fact that weather cannot be effectively predicted but is a major source of crop yield variation, is a severe limiting factor to DSSs, since the success of many systems depends on their ability to predict future weather.

11.6 Criteria for Success of Decision Support Systems

Hamilton *et al.* (1991) suggest that, despite the poor record models have of being applied, optimism on the part of their developers about their future potential is not flawed, but based on their realization that there now exists a tool for complex decision making in a way that was not possible previously. This potential is only just beginning to be realized. Many of the reasons for poor adoption may not reflect poor potential. The following issues should be considered in relation to the potential for successful model application in LDCs.

It is important to consider what constitutes success in terms of model development and application. Zadoks (1981) believed that, when EPIPRE made itself superfluous by furthering farmers expertise, it had then done a good job. Others clearly believe that if DSSs have made themselves obsolete, then they have failed. Newman *et al.* (2000) question the way in which success is defined:

- Is a DSS successful when it matches the decisions made by experts?
- Is it successful when it reaches a certain percentage of the target market?
- Is it successful when it continues to exist and has some use amongst researchers, extensionists and farmers?
- Is it successful when it becomes obsolete through use of the decisions it has made?
- Is it successful when farmers, through some other form of media, use the information generated to make tactical or strategic decisions?

Defining success from the user's perspective may help in deciding whether a model is an appropriate tool to use in a given situation. Success from the user's perspective may be different from that of the model developers, or a company trying to market the software. For example, scientists may consider a model to be successful when it provides them with useful information about the impact of climate change, however, if this information is not used to inform decision making, then the model output will not have served a practical purpose. If models are to be useful in a development context, then their success should ultimately be measured by the impact that they have on farming systems and natural resource management.

11.7 Risks Associated with using Decision Support Systems

Many commentators observe a number of underlying risks associated with model use which must be considered when implementing models for decision support.

1. *Misuse*: Philip (1991) suggests that models can lead to decision makers avoiding responsibility by using the model as the truth to back them up (i.e. the model says it, so it must be true). He considers the risk of unscrupulous decision makers seeking modellers who will design a model to give the answers that they want. Although this fear is certainly valid, corrupt decision makers will probably find means to further their personal agendas even in the absence of computer technology.

2. *Simplification leads to misleading outputs*: There is a danger that simplified models could suggest paths of action that are inappropriate, since models often only contain the components that the user thinks are important (Spedding, 1990). For example, traditional ways of modelling whole farms have often focused on the economic optimization, and have not generally considered wider costs to society, e.g. pollution.

3. *Confusion between model and reality*: There is danger that users will not be aware of the assumptions and limitations of the software that they use, and will confuse the model with reality (Philip, 1991).

Given the low level of uptake of DSS in agriculture at either the tactical or strategic level, however, we may sleep easily in the knowledge that it is unlikely that Philip's (1991) fears will be realized. Of far greater concern is that the immense effort and investment in DSS research appear to be foundering on the divide between theory and practice. McCown (2001) advises that 'the successive shifts in concepts toward "softer" modes of intervention are stages in painful realisation by scientists that formal models, although sometimes valuable in research practice, simply are not appropriate to the physically and socially situated practice of farming.'

Whether it is possible to disseminate widely enough the necessity to bridge the gap between science-based DSSs and farming practice remains to be seen. The record to date suggests that it will be some time before there is true collaboration between researchers and farmers involving DSSs, especially in developing countries.

Part 3
Models as tools in education and training

Using Models as Tools in Education and Training

12

Anil Graves, Tim Hess and Robin Matthews

Institute of Water and Environment, Cranfield University, Silsoe, Bedfordshire MK45 4DT, UK

12.1 Introduction

The potential of computer simulations in education was increasingly examined in the 1970s and has now become common practice in many developed countries. The role of computers is becoming increasingly important as prices fall and computing power increases, which doubles roughly every 18 months according to 'Moore's Law' (Kaku, 1998). Because of this, and the increasing global demand for a computer literate workforce, the question is sometimes raised as to whether computer-assisted learning (CAL) should be used at all in education and training in less-developed countries (LDCs), is perhaps superfluous. CAL is probably inevitable, especially in science, and there are good reasons for this. As Jeger (1997), suggests, '. . .computer education has become inevitable in dealing with the full range of complexity inherent in agriculture in general.' Narasimhan (1995), notes that models of some description are 'intrinsic to the scientific method' and that they 'help to recognise patterns in the structure of nature', with the patterns thus recognized used to 'extrapolate' and 'predict'. Furthermore, 'within the bounds of its role in the scientific framework, the computer can help evaluate complex interactions and visualise complex objects in ways that have never been possible with the classical analytical methods of solution'. Thus, although many LDC educational institutions may not presently have the resources to use crop–soil simulation models in education and training, attention should begin to focus on how best such models could be used in this context, rather than on whether they should be used at all.

The discussion presented here has been developed from literature describing the use of crop–soil simulation models as educational tools in developed countries, as well as in LDCs. This is largely because the pool of literature available on the use of simulation models in education and training in LDCs, is relatively small. The wider experience of simulations and CAL tools used in other subjects has also been important here, in developing an understanding of how best to use crop–soil simulation models for education and training. There are important questions that need to be answered with regards to the use of crop simulation models in this way. For example, where can simulation models contribute most to education and training? When and how should they be used and not be used? What are the desirable software features for such models? By drawing attention to the positive and negative impacts of crop simulation models used as tools in education and training, we hope to be able to provide a rudimentary framework for good practice.

The use of crop–soil simulation models is seen here in relation to their use in further and higher education. At school level in LDCs, the use of such tools is generally not practical, due to the limited computer resources available. Nor is it necessarily desirable, as many of the lessons that can be drawn out of crop–soil simulation models are most useful at a more advanced level.

Young and Heath (1991) observed that simulation models are the 'most widely accepted approach to CAL focusing in the main on problem based learning'. They suggested that two approaches could be used.

1. *Experimentation and observation using existing simulation models*: The crop–soil simulation model may be used in place of conventional field or laboratory experimentation.
2. *Model building*: The process of defining and building a crop–soil simulation model and its implementation on a computer.

The approach adopted will depend on the nature of the lesson being delivered to the students, the approach of the tutor, and the resources of the educational establishment. Within the context of the present study, the emphasis is on the application of existing stand-alone crop–soil simulation models in the area of experimentation and observation. Although peripheral to the main discussion, the benefits of model building in education and training will also be discussed briefly.

12.2 Using Existing Simulation Models

The primary role of crop–soil simulation models in education has been as a substitute, or adjunct, to traditional field or laboratory experiments. There are two main applications of simulation models in education and training:

1. Training to perform real experiments or tasks. The classic example of this is flight simulation.

2. Replacement of some real world experiments with computer simulations.

In the case of crop–soil simulation models, the first of these is rarely appropriate – however, simulation models are commonly used as a substitute for real-world experiments. The student can 'plant' a particular crop in a chosen soil type, subject it to particular environmental influences and study its performance. The student can then simulate different 'treatments' to investigate the impact of these on plant growth and yield.

12.2.1 Benefits of crop–soil simulation models

There are benefits of such an approach, to the student, the tutor and the educational establishment.

1. *Speed*: A major limitation of traditional field or laboratory experimentation in plant science and agricultural studies is the time that it takes to grow a crop. A 3-month growing season is often incompatible with the time frame of students' courses. Therefore, while such experiments feature in research programmes, they are rarely incorporated into teaching programmes. A model allows the simulation of the complete growing period (or even of several seasons) in a matter of seconds. Crop–soil simulation models are therefore well suited to answering 'What if . . .?' questions and a greater range of treatments can be examined. This allows for more effective use of learning time and the chance to discover by trial and error.

2. *Controlling the 'environment'*: Experimentation through simulation can also allow the student to control the 'environment' and isolate the influences of certain input variables. Field experiments are often confounded by uncontrollable and unpredictable environmental influences (e.g. weather, pests and diseases), which may produce 'bad' results and obscure the anticipated learning outcomes. The use of a simulation model allows environmental factors to be 'controlled' so that the impact of the treatment under consideration can be isolated and scrutinized.

3. *Performing impossible or undesirable experiments*: Crop–soil simulation models allow students to study the impact of perturbed environments that are impossible or expensive to create, such as higher temperatures resulting from climate change. The outcome of experiments which may be undesirable or unethical in real life, such as the impact of pollution, can also be examined (Centre for Land Use Studies, CLUES, 1996b). In the life sciences, McAteer *et al.* (1996) found simulation models useful in avoiding ethical difficulties, such as those relating to practical experimentation on live animals. An interesting example of a simulation that can be used to avoid such ethical difficulties (and save the lives of frogs!) is the

'Virtual Frog Dissection Kit'.[1] The aim of this kit was to provide school biology classes with the ability to explore the anatomy of a frog by using graphical computer simulation to visualize the anatomical structures of the intact animal (Robertson *et al.*, 1995).

4. *Providing a safe environment for experimentation*: Simulation models used as a substitute for practical experimentation provide a safe working environment, and do not require high levels of 'physical or technical skills' (McAteer *et al.*, 1996). Pilkington and Parker-Jones (1996) also noted that '. . . simulations offer a safe environment in which to practice making real-world decisions' allowing tasks to be set 'when it would otherwise be not safe to do so'. This can be important in education and training, as it allows students to make important or dangerous decisions in the 'virtual' world, which have no direct effects in the real world.

5. *Focus on the subject matter*: In the life sciences, McAteer *et al.* (1996) compared the use of simulation models as an alternative to traditional 'wet labs'. They found that students spent less time learning how to use computers and software than on learning how to use laboratory instruments and equipment. This allowed students to spend more time focusing on the subject matter itself, rather than on the techniques required for real-world experimentation and observation.

6. *'Observing' obscure processes*: Many biophysical processes cannot be effectively observed in the laboratory or can only be observed using expensive equipment. A simulation model can clearly demonstrate those processes to the student, allowing greater insight into the 'causes' and 'effects' that are simulated, than would be possible in simple observational experiments (McAteer *et al.*, 1996).

7. *Substituting for time*: Teaching staff in educational institutions are often stretched for time, and the use of crop–soil simulation models as part of a CAL programme is seen as a means of redressing decreasing staff/student ratios (Young and Heath, 1991) and declining contact time (McAteer *et al.*, 1996). They are also seen to be a way of substituting for the tutor, in the more clerical aspects of individualized learning programmes, giving them more time to concentrate on other university work, such as research (Jovanovic *et al.*, 2000).

8. *Substituting for financial and physical resources*: Physical and financial resources, such as laboratory equipment or land on which to conduct experiments, may also be increasingly stretched in educational establishments. Simulation models allow lecturers to 'extend their students' experience in a climate of diminishing resources and increasing student numbers' (CLUES, 1996a), and overcome the need for multiple sets of specialized or expensive equipment (McAteer *et al.*, 1996).

[1] http://www-itg.lbl.gov/vfrog/

9. *Documenting and transferring research expertise and project experience*: Boote *et al.* (1996) suggest that crop simulation models are effective in documenting the experience of research experiments and projects; 'the data files become permanent documentation of experiments which are available to others rather than becoming lost in filing cabinets'. Crop simulation models can also embody the expertise of scientists gained over many years at the forefront of such research. Such expertise and experience can often be lost and may be difficult to find, use and transfer to students, if traditional paper-based methods are the sole means of documentation.

10. *Elucidating complex relationships and interactions through dynamic simulation*: The amount of information generated by scientists on crop growth and crop–environment interactions is often difficult to convey to students through traditional means (Simmonds *et al.*, 1995). The Computers in Teaching Initiative (CTI) suggests that 'one of the best ways to promote deep conceptual understanding of the real world is through the investigation of simulation models' (CLUES, 1996b). Crop simulation models provide a dynamic representation of these complex interactions, giving tutors an additional dynamic means of elucidating complex relationships and interactions for students.

11. *Synthesizing fragmented knowledge*: Due to the tendency of the scientific process towards reductionism, the amount of scientific knowledge generated in the plant sciences is possibly too large to be usefully appreciated, except through simulation models. Crop simulation models provide a way of synthesizing fragmented research understanding. As Boote *et al.* (1996) suggest, 'Crop models are particularly valuable for synthesising research understanding, and for integrating up from a reductionist process.'

12. *Synthesizing diverse disciplines*: Boote *et al.* (1996) have noted that models are 'valuable interdisciplinary research tools that integrate discipline knowledge and relationships to produce a descriptive tool for application beyond the individual science discipline'. The need to synthesize diverse disciplines for effective real-world impact is increasingly important, as the growing importance of multidisciplinary research suggests. This is particularly evident in research for LDCs, where a socio-economic as well as a technical understanding of agriculture is often required (see Chapter 15). This integrated understanding is also increasingly important for students, who will become researchers and managers of environmental systems in the future. With reference to crop protection, for example, Jeger (1997) makes the point that more than an understanding of hard facts is required for crop protection measures to be effective. Miller (1993) suggests that agricultural and environmental issues are often 'trans-science' at the decision-making and policy level. Crop simulation models may provide a means of integrating across traditional discipline boundaries, or may provide a component in a larger simulation that does so.

13. *Gaming*: Simulation models are frequently used in problem-based learning exercises where students can investigate a model as a substitute for the real world. However, the models can also be used in the form of a 'game' in which the students are required to find an optimum set of inputs to achieve a desired outcome (CLUES, 1996b). Through trial and error, the student can develop a set of heuristics by which to predict the impact of changes in inputs on the final output, and gradually 'home in' on the optimum solution. In this case the model may be used as a 'black box' as an understanding of the structure and mathematics of the model is not required. Crop–soil simulation models may also be a component of larger simulation games. Such games, for example, are used to deliver ongoing training to managers of irrigation schemes (Dempster *et al.*, 1989). Edwards (1997) has noted that whole-farm simulations force students to integrate economic decisions with financial management decisions. In gaming scenarios such as these, Burton (1989) has also observed that the simulation can become the focal point around which students' wider experiences become part of a collective learning experience.

14. *Distance- and self-learning*: Distance-learning has been growing in developed countries and its importance is also likely to increase in developing countries, as on-the-job training and adult education increases. Many students may be unable to attend formal learning centres for practicals and laboratory experiments. Edward (1996), for example, reports on how a simulation model was used in place of laboratory experimentation for offshore oil workers studying on undergraduate programmes. The learning experience of students undertaking distance-learning courses can be greatly enhanced with CAL providing a complement to textbooks. The British Association for Information Technology in Agriculture (CLUES, 1998) is even more optimistic about the role of information technology (IT) in agricultural education and foresees a future of continuous self-education with web-based learning tools and CAL software as the normal state of affairs for adults wishing to update their agricultural knowledge. This is of little relevance to the third-world farmer directly, but extension workers and researchers in LDCs could find such self-education increasingly important in the future. Practical experimentation or field observations for students learning at a distance is not always possible. This has stimulated the use of crop simulation models as an educational resource to be used in the physical absence of the teacher. As an example of this, Ortiz (1998) describes how the SOYGRO model was adapted for use as a distance-learning tool in Costa Rica.

12.2.2 Examples of crop simulation models used as tools in education

TRITIGRO

TRITIGRO I (McLaren and Craigon, 1981) was used to teach students about the impact of husbandry decisions on whole crop growth. Although the classic process of science involves experimentation and observation of the real world, they found that this was not always possible, given lack of facilities and finances. TRITIGRO I was developed to allow students to undertake 'virtual' experiments in the absence of these resources. The responses built into the model were based on a combination of published responses, experimental facts and practical experience. As such, McLaren and Craigon (1981) point out that some of the responses of the model were subjective, and that the model would also have been limited by the knowledge and ability of the developers. However, they point out that these same limitations would have applied with other forms of teaching. A synoptic outline of the programme was presented, explaining how students could select establishment and management strategies and how this would affect whole-crop responses. For example, students could select the variety of wheat to be sown, the seed rate and the sowing method.

The application was also used to assess the students, the idea being that students familiarize themselves with and consult the relevant literature and lecture notes, before and while undertaking the assessment. Although the assessment would have had built into it the subjectivity of the lecturers, McLaren and Craigon (1981) point out that the students could all be assessed with exactly the same degree of subjectivity.

McLaren and Craigon (1981) felt that the crop simulation model was able to deliver many important management lessons to students. For example, the model was able to demonstrate that, rather than the total amount of money spent on the crop system, it was targeted spending on appropriate management factors that delivered better results. Students who did not spend a relatively small amount of money on pest control during the establishment of the crop found that the penalty for this multiplied later on, as yields were reduced. Even large amounts of spending on fertilizers and fungicides could not later make up for this management error. Lessons and experiences that might otherwise take months to learn through real world experimentation and observation could be dynamically demonstrated in a matter of hours with the crop simulation model.

A questionnaire survey undertaken by the authors showed that, in general, the students felt their interaction with the crop simulation model had been beneficial. Many of the students felt that the crop simulation model should be expanded and that the development of models for other crops would also be beneficial. Much thought and discussion of inputs, as well as of results, seems to have been stimulated by the crop simulation model, an occurrence that the authors felt was very positive. About 80% of students felt that they had gained more benefit from their interaction with the crop simulation model than they would have had from an essay-type project.

SOYGRO

Ortiz (1998) describes how the soybean simulation model, SOYGRO V5.41, was adapted to allow students in the school of Natural and Exact Sciences at the Universidad Estatal a Distancia (UNED), Costa Rica, to successfully complete modules in the Agricultural Business Administration Programme. The distance-learning methodology employed in this programme breaks the time and space barrier to education, allowing students to learn in any part of the country at any time, even while still working. SOYGRO was used to simulate the effect of soybean growth, along with multimedia packages, written materials, audio-visual materials, study guides and tutorials.

Various difficulties had to be overcome during the development and testing of SOYGRO as a teaching tool. These ranged from computer hardware problems to computer literacy problems. An introductory program was written in BASIC to facilitate the students' interaction with the computer. Despite these initial teething problems, SOYGRO was successfully implemented as a mandatory practical in the Basic Grains course from 1992, 'as a support element in the theoretical aspects of the course', and has gained widespread acceptance as a teaching tool in distance education (Ortiz, 1998).

Ortiz (1998) shows that there were several important steps required before SOYGRO could be effectively used in distance education.

1. *Testing and evaluation*: The first step was to test and evaluate the model and its functions. This was done with university students from the agricultural and computer science courses. The course guide and envisaged teaching process were revised on the basis of the students' experiences and questionnaire surveys.

2. *Adaptation of SOYGRO to a new physical context*: The second step was to provide information required to run the model for Costa Rican conditions. Three minimum data sets for Costa Rican conditions were found from the Centro Agronomico Tropical de Investigacion y Ensenanza (CATIE).

3. *Adaptation of the users' guide to a new educational context*: The third step was to adapt and create a users' guide for distance education in Spanish. This was based on the original Users' Guide. Two practical exercises were also developed for the students.

4. *Staff training*: The fourth step was the development of a training programme for the staff of the Natural Resources and Exact Sciences School on the use of the model and its Users' Guide.

The process was not without teething problems, but the experience was positive enough to encourage the development of a simulation model for banana diseases in the Fruit Crops course at the same institute. It was hoped that this would help to give students an understanding of management strategies that can counter the development of diseases in bananas.

PARCH

PARCH (Bradley and Crout, 1994) was developed at the University of Nottingham in a Department for International Development (DFID)-funded project, and simulates the growth and development of sorghum, millet and, to a lesser extent, maize, in semiarid environments. The original purpose of the model was to aid in the analysis and extrapolation of experimental results in Botswana. However, the attention given to the development of the user interface makes it suitable for use as a teaching and learning tool for those studying tropical agriculture and agronomy, and helped to transfer the research experience of the scientists in the semiarid tropics to students in other areas. This is important, given the difficulty of visiting and conducting real experiments and observing real crops in other countries, within the short time frame of a university course. Fry (1996) suggests that the 'transparency of the interactions between crop, soil and climate within the model were considered to make PARCH an ideal teaching tool'. He goes on to suggest that it is 'most readily used as a teaching tool, both to illustrate modelling concepts and to show the complex plant–soil–atmosphere interactions in agricultural environments; situations where great accuracy in the inputs and outputs is not essential'.

For example, PARCH could allow students to examine how light and water are 'captured' on a daily basis by the crop and converted into assimilated dry matter. Students can further examine how crop growth is limited by either light or water and relate this to the crop's stress response (such as leaf rolling or increased partitioning of roots). An important component of PARCH is a water balance simulation and this is an essential component in determining final dry matter production. Students are able to examine how water demand is driven by intercepted solar radiation and examine how this links in with water availability and the final yield of dry matter. A tutorial guide has been written for PARCH which aims to draw out specific lessons (Stephens *et al.*, 1996). This leads the student through pre-prepared exercises, and encourages them to consider and to give reasons for the results they obtain. PARCH has been used by staff at Cranfield University to support courses in crop physiology. In a study of the uptake of PARCH, Fry (1996) found that it was being used by several university lecturers to support their courses.

SPACTEACH

SPACTEACH (Simmonds *et al.*, 1995), was originally constructed as a research tool, but was further developed to use as a CAL tool to enable students to examine water movement in the soil–plant–atmosphere continuum. The main impetus behind the development of SPACTEACH was the difficulty encountered in conveying concepts with a 'blackboard full of equations' (Simmonds *et al.*, 1995). The software was seen as a way of bringing to life the complex dynamics of water movement in a soil–plant–atmosphere continuum. The movement of water in such a continuum is a key function of the availability of water for crops, the movement of pollutants through soil, groundwater hydrology, energy budgets of land surfaces and erosion caused by runoff. The students can experiment with the effects of changing vegetation cover, rainfall and soils. SPACTEACH can, for example, be used in a problem-solving or investigative context, and teaching materials have been developed to help fulfil this. The learning experience is improved by the use of a Windows interface, and simple and colourful graphical outputs.

The main objective of the development of SPACTEACH was to provide the teacher with an additional teaching tool rather than to provide a substitute for the teacher, and lecturers are encouraged to develop their own materials to be used with SPACTEACH. However, a Lecturers' Manual is also provided which gives examples of how to use SPACTEACH and how to draw out the salient lessons that the model can provide. Such documentation is important as it allows tutors to integrate crop simulation models more easily into their existing courses.

PLANTMOD

PLANTMOD[2] was developed for use both as a research and a teaching tool. Its major aim was to demonstrate five important processes – plant growth, light attenuation through a canopy, photosynthesis, transpiration and canopy temperature. Each of these processes is graphically demonstrated and the user can manipulate various physiological and environmental inputs to examine their effect on the five processes noted above.

The use of the model and its outputs are made simple with a Windows-style interface and easy exporting facilities for graphs and results. The findings can therefore be written into essays and reports using a standard word processor. According to Batchelor (1997), who reviewed the model, the manual is clear and easy to use, with individual chapters describing the mathematics behind each of the five processes. Students can, for example, examine plant growth through the logistic and Gompertz growth functions. Sensitivity analysis can then be conducted on the predicted growth rates and the initial and final plant weights for each function. Similarly, students can examine light attenuation using Beer's Law, photosynthesis in C_3 or C_4 plants, and the environmental and physiological factors affecting transpiration using the Penman–Monteith or Priestly–Taylor equations.

---

[2] http://www.greenhat.com/plantmod.htm

Batchelor (1997) found the model to be an excellent tool for teaching several plant physiological processes. In his own classroom tests, he discovered that students found the software easily accessible. It elucidated the five processes covered above, by allowing students to see what happens to them when various environmental and physical characteristics were manipulated. Students could also undertake trial and error experiments and develop 'What if . . .?' scenarios. He suggested that PLANTMOD could be used as a supplement to normal teaching materials in university plant physiology courses, and that it could easily be used for assignments and laboratory exercises to reinforce classroom teaching.

Soil water balance (SWB)

SWB is a 'mechanistic, generic crop irrigation schedule model, making use of weather, soil, and crop databases to simulate crop growth and soil water balance' (Jovanovic *et al.*, 2000). Jovanovic and Annandale (2000) described the use of SWB for irrigation management training in South Africa, where critically low water supplies in rural communities are likely to be further stretched by growing populations and per capita demand for water. The aim of using SWB in this context was to familiarize students with the use of computers for problem solving and to allow them to 'experiment with various input parameters to determine which would have the greatest effect and to provide problem-solving management experience' (Jovanovic and Annandale, 2000). Jovanovic *et al.* (2000) described its use in crop physiology courses, where it was used to demonstrate the impact of varying a number of inputs on the growth and productivity of irrigated maize. Questionnaire surveys of students by Jovanovic and Annandale (2000) and Jovanovic *et al.* (2000) showed that students felt that they generally benefited from using SWB.

Jovanovic *et al.* (2000) and Jovanovic and Annandale (2000) outlined useful procedures for the use of simulation models derived from their experience of the use of SWB. Importantly, this involved students developing a proper understanding of the theoretical background to the processes being illustrated by the model, before demonstration or use of the model took place. Well-developed assignments, effectively linking what was presented through traditional teaching materials and the crop simulation model, could then help to increase the students' understanding of the issues involved. Further innovations include the suggestion that simple simulations could be used to illustrate individual components of the resource system, such as infiltration, drainage, canopy interception, evaporation and transpiration, after each topic has been covered theoretically in class. Only then would the dynamic interaction and complexity of the resource system be illustrated with the aid of a complex mechanistic model (Jovanovic and Annandale, 2000). The practicality of this would have to be considered in terms of model features, development and availability.

12.2.3 Comparison of crop simulation models with the traditional educational experience

The assessment of crop simulation models as educational tools tends to be anecdotal in nature, and the debate typically revolves around the issue of whether students have learned anything at all through their use. The question as to how effective crop simulation models have been as an alternative way of transferring information, compared with more traditional forms of teaching, has mostly been ignored. CLUES (1996a), for example, reported that the implementation of simulation models in two British university degree programmes increased motivation and allowed the students to develop a deeper understanding of the underlying principles of the subject than was possible with previous approaches, but did not use statistical evidence to show this.

Although there is little peer-reviewed literature testing the statistical differences between students taught entirely through crop simulation models and those taught through traditional means, attempts to quantify the impact of computer simulation on student learning have been made in other subjects. Coleman *et al.* (1998) compared a group of computer-engineering students learning entirely with CAL applications with a control group taught through traditional lectures. Tests indicated that there was no significant difference in the performance of the groups subjected to the two methods of teaching. Edward (1996) compared the learning achievements of two groups of students undertaking HND and BSc engineering courses – one group using a conventional laboratory and the other group using a multimedia package, including a simulation model. Learning was determined by tests before and after the use of the learning materials, a report required for 'summative assessment', and by student responses in questionnaires and interviews. Edward (1996) concluded that learning was at least as effective with the computer-based package as it was with the conventional laboratory approach.

In a study of a large pharmaceutical company, Williams and Zahed (1996) compared staff trained using traditional lectures and those self-taught through computer modules. Both groups demonstrated significant learning after training and there was no significant difference in the level of learning between the two groups. Interestingly, the self-taught CAL group retained significantly more information after 1 month than the traditionally taught group. There was, however, no significant difference in satisfaction with the training experience.

Forsythe and Archer (1997) found in a correlation study that psychology students rated the use of CAL tools positively, and that 'those with weak academic backgrounds who consistently used the technology achieved higher test results than weak students who did not use the technology'.

These results suggest that CAL can be as effective as traditional teaching methods, and may in some cases have positive benefits in terms of motivation and retention. However, it should be noted that there are also negative aspects to the use of CAL tools, which we discuss later.

12.3 Model Building

Although model building is peripheral to the main theme of this book, i.e. the application of existing crop–soil simulation models in developing countries, it is worth briefly noting some of the advantages of the model-building approach. There are various benefits to be derived from model building in education and training, most of which stem from the more active participation of the student in the learning process. Teaching model-building skills also develops modelling capability, which may be important in localized research and extension situations, where models do not already exist or where existing models are felt to be unsuitable in some way. The benefits of model building are shown below.

1. *Building models for experiential learning*: The development of the model is in itself an active and creative process. Students are active participants in their own learning rather than passive recipients of information through traditional means. Although the use of crop–soil simulation models for experimentation and observation involves greater participation from students in the learning process than the use of paper-based materials, the process of model building takes the process of participation one stage further, by allowing students to create the models themselves. The process involves setting a systems problem for the students. They can gradually build up the complexity of the model by adding more and more components to their system. This can be done in conjunction with their study of other paper and electronic materials.

Building computer models has been greatly aided by the availability of software that allows them to be developed visually, rather than with programming code (e.g. MODELMAKER[3], STELLA[4], SIMULINK[5]). These allow students to 'draw' their models on-screen, link and define mathematical relationships between components, run the model, and graph results without the need to learn code. Thus the assumptions and formulae underlying the processes of the model are easily examined and changed by the student. The use of graphic modelling software not only allows easy access to the principles embodied by the model, but also allows people with practically no programming skills to develop their own models. Such tools have proved extremely useful in teaching modelling principles and applications in environmental and agricultural sciences (Morison,

[3] FamilyGenetix Ltd, UK. http://www.modelkinetix.com/
[4] High Performance Systems Inc., USA. http://www.hps-inc.com/
[5] The MathWorks Ltd, UK. http://www.mathworks.co.uk/

1995). Evidence suggests that because students gain the knowledge and understanding required for the subject in a highly engaging manner, they assimilate the material more quickly and retain it for longer. Students also develop a set of critical thinking skills – specifically those related to 'systems' thinking.

2. *Developing future modelling capability*: Modelling capability can be an important skill for those working with natural resource systems, and in theory such capability should equip students with the ability to develop models for a variety of systems during their future professional lives. Simple functional models in particular could be developed by professionals in LDCs to account for local contextual situations. This might be especially useful where existing simulation models are not available for a local crop, or where they have not been calibrated and validated for local conditions. This could provide a certain degree of independence in the future, but would necessitate computer literacy and computer resources, both of which may be lacking in certain LDCs. Although model building has been made easier by graphical modelling environments, there is no antidote to lack of funds for resourcing. Additionally, complex problems might still require an expert system and the involvement of professional programmers.

3. *Building models to elucidate complex relationships and interactions*: Although the use of pre-existing crop simulation models might allow complex relationships and interactions to become evident to students, these processes are likely to become clearer and more likely to be retained when they have developed their own models to solve a specific problem. The exact extent and importance of these relationships become clearer as the students develop and experiment with the output from their own models.

4. *Building models to synthesize fragmented knowledge*: The synthesis of fragmented knowledge is increasingly important in science. The reductionism that is part of the scientific process can make scientists oblivious to the larger picture. In the same way that crop simulation models can synthesize fragmented knowledge, building models can also allow students to appreciate the larger picture. Active participation in the creation of a 'system' may once again make this lesson all the more evident.

5. *Building models to synthesize knowledge across diverse discipline boundaries*: The active process of synthesizing information from different disciplines ensures that the student is aware that cropping systems operate in a wider context. Existing social and economic systems are important in determining the success of cropping systems interventions and the sooner the student understands this, the better. This wider perspective may be well developed in large simulation games, but the creation of a simple farming system may impress upon the student the importance of a perspective that integrates various disciplines.

6. *Identifying knowledge gaps*: CLUES (1996a) reported how the 'Carbon Turnover in Soils' CAL package has been used to allow students to test their simple spreadsheet models. They soon discovered that the simple one-pool first-order kinetics model does not explain the trends in the data. The students then had to modify their own model into a two-pool model. Such an appreciation of the 'knowledge gaps' that might exist in models should help to instil a healthy understanding of the possible limitations of models and ensure that students do not simply accept modelling outputs at face value.

12.4 Transferring the Systems Approach to Less-developed Countries

There are two major pathways by which crop simulation models may come to be used as tools in education and training in LDCs. The first of these pathways is largely demand-led, where users in LDCs decide to use or develop their own models, either for their own learning or possibly as teaching tools. This pathway tends to be informal and unstructured where users select the model they consider to be best for their particular purpose. In the second pathway, crop simulation models can be transferred to LDCs through a structured route. This approach tends to be supply-led, and largely initiated by institutions in developed countries, ultimately also supplying the technology and training (Fig. 12.1). The use of crop simulation models through the supply-led route have tended to focus on developing and then disseminating pre-existing crop simulation models, as, for example, with the IBSNAT project (Tsuji, 1993), or to focus on developing the capability to build crop simulation models, as in the SARP project (ten Berge, 1993). Amongst the various aims that are cited for such projects was the stated need to develop a systems analysis approach to natural

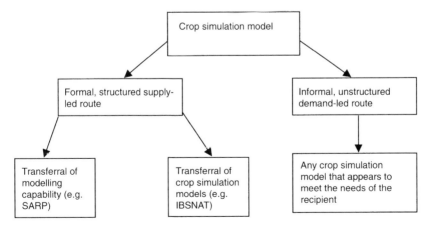

Fig. 12.1. Schematic representation of possible technology transfer routes.

resource research and management. The use of crop simulation models and the development of modelling capability have been seen as a possible vehicle for this process.

12.4.1 The IBSNAT project

The International Benchmark Sites Network for Agrotechnology Transfer (IBSNAT) was initiated in 1982. The major aim was to allow scientific knowledge to be used by the non-scientific community, particularly farm advisers and policy makers, by embodying the information in crop simulation models (Tsuji, 1993). Because of rapidly changing agricultural technologies and the urgency of the perceived need in LDCs, technology transfer was thought to be a major possible option for providing food security in LDCs. The belief was that agrotechnology transfer could be more efficiently made through systems analysis and a global network of scientists.

Several crop simulation models already existed at the time of IBSNAT's inception, and one early objective of the project was to enhance these models so that they could simulate the outcome of alternative crop production strategies anywhere in the world. In this way, the selection of a specific technology for use in new cropping systems could be rapidly demonstrated at no cost to the technology adopter. The need for time-consuming on-farm trials would be minimized and 'What if . . .?' questions could easily be answered. Optimism was particularly high at this time, due to the success that the United States Department of Agriculture had previously in predicting the yield of maize in countries where the model had never been used before.

The main product of IBSNAT was the Decision Support System for Agrotechnology Transfer (DSSAT) software package. This consisted of:

1. A database management system to store and retrieve the minimum data-set of soil, crop, weather and management data to validate and apply the crop simulation models.
2. A set of validated crop models to simulate the outcomes of genotype × environment × management interactions.
3. Application programs that facilitated the manipulation of the databases, the use of the crop models, and the presentation and analysis of the model output.

Great effort was made to ensure that the models could be easily used and this was instrumental in developing the concept of the minimum data set (MDS). The main aim of this was to ensure that the data requirements of the models within DSSAT were not so large as to deter people from using them. The DSSAT shell enabled a variety of different models to be accessed through a single user interface.

According to Tsuji (1993), DSSAT models have been widely adopted by various users in various capacities. DSSAT was also integrated into a spatial simulation program called AEGIS (see Chapter 6 for more details). The success of IBSNAT can be measured in terms of the uptake and use of the DSSAT software package – the estimated number of users grew from about 50 in 1989 to over 500 in 1992, representing the recorded number of distributed copies of Version 2.1 of the software. By 1994, 300 more users were using Version 3. These users were located in about 80 different countries.

The fact that so many people were involved was a defining element of the project (Tsuji, 1993). Membership of the IBSNAT network was open to any individual or organization willing to share data, information, models and experiences. The value of these contributions was estimated to be about six to seven times what was given by USAID as seed money (Uehara and Tsuji, 1993). The lessons and products of IBSNAT were disseminated by technical publications, technical and research reports, user guides, conferences and symposia proceedings, newsletters and progress reports (Tsuji, 1993). Another essential element of the technology transfer process was the training courses. Tsuji points out that over 600 people have been trained by IBSNAT workshops and training sessions in various parts of the world. It is likely that many more people have been trained to use DSSAT by other organizations, or have taught themselves how to use the models.

Tsuji (1993) makes the point that the project in itself was a global endeavour, and that the product was greater than what any individual organization could have produced alone. Systems simulation was a recent advance in agricultural science in the 1980s and the expertise to develop a tool like DSSAT could only have been possible with multidisciplinary effort and global expertise. A strong network of users still maintains contact through a dedicated list-server which provides support both from developers and from other users, even though the project itself finished in 1994.

Independent assessments of IBSNAT and its impact on decision making in LDCs appear to be rare. However, the use of DSSAT models in research is widespread, and it can only be assumed that such use forms the basis for decision making, several links away from the decision maker (see Section 9.5 for an example). The models within the DSSAT shell have been applied to a wide range of important issues ranging from the impact of global warming on crop production (Luo *et al.*, 1995; Buan *et al.*, 1996) to the use of DSSAT models in education (Ortiz, 1998). Tsuji *et al.* (1998) have compiled a large amount of literature on IBSNAT that also demonstrates the wide variety of research problems tackled by users of DSSAT. Although it is difficult to find independent evaluations of the IBSNAT project, there are clearly important lessons to be learned from the IBSNAT experience and some of these may be relevant in the future if attempts to develop and devolve crop simulation models to LDCs continue to be made:

1. *Extensive preparation*: Preparatory stages were undertaken to ensure that the products of the project would be relevant to end-users. These fed back into model and database design.

2. *Minimum data requirements*: The concept of the minimum data set was established to ensure that users of DSSAT would not be swamped with huge data demands.

3. *Multidisciplinary approach*: The scope of the project was large and no single institution would have been capable of developing the DSSAT software alone. A multidisciplinary team was assembled instead.

4. *Global resourcing*: The global scientific resourcing of the project was important for the development of DSSAT.

5. *Effective networking*: Effective networking has proved to be important for the continued use of DSSAT. Solutions to problems can be offered to the user both by model developers and by other users through list-servers.

6. *Effective dissemination*: The dissemination of DSSAT was one of the primary objectives of the IBSNAT project. This was done through journal publications, reports and training workshops.

12.4.2 The SARP project

In 1984, funding was obtained from the Dutch Government for the establishment of the Simulation and Systems Analysis for Rice Production (SARP) project, to be run by Wageningen Agricultural University (ten Berge, 1993). The project was conceived with the objective of creating more productive and sustainable agriculture by improving agricultural research, extension and planning in developing countries for rice-based cropping systems, using a systems analysis approach. Crop simulation models were considered to be important for increasing the efficiency of research by identifying research priorities and reducing the costs of experiments. It was perceived that issues of environmental importance and sustainability could also be tackled through systems simulation. The International Rice Research Institute (IRRI) in the Philippines coordinated the project, and strengthened links between IRRI and National Agricultural Research Centres (NARCs), to transfer the required skills and training to multidisciplinary teams in each NARC. Further support came by conducting collaborative research between IRRI, NARCs and Wageningen University, which would continue to bolster the use of the systems approach in NARCs. Support was to be given to NARCs undertaking to train additional staff in the use of the systems approach and undertaking to coordinate future research activities. The eventual aim was to stimulate locally coordinated programmes of research in rice systems, supported by the capacity to generate indigenous crop simulation models. An important feature of the SARP approach was that models were not presented as final or unalterable products. Rather, the idea was to understand the principles of modelling and to allow users to adapt the models to their specific needs (Goldsworthy and Penning de Vries, 1994: 334).

The project consisted of three phases. The first ran from 1984 to 1987, the second from 1987 to 1991, and the third for a further 4 years until 1995. The focus in the first two phases was on training of members of the collaborating teams in systems analysis techniques, while that in the third phase was on applying the models to topics proposed by the teams on the basis of their relevance to local conditions and the ability of the models to deal with the problems identified. Three training programmes were executed in the first two phases, each consisting of three components.

1. A basic training in simulation and systems analysis (6–8 weeks).
2. A case study conducted by each team at their home institute (8 months with a 1-week visit by SARP staff).
3. A workshop where findings were presented and new areas of research identified (2 weeks).

The SARP project identified various factors that would facilitate the transferral of the simulation approach and modelling capability to rice systems research. The intertwining of training and research and the joint collaborative research approach was seen as essential to the process of transferring modelling skills and systems tools. The identification of multidisciplinary interaction among scientists at the NARCs was also important. The development of specialized software tools and models that were transparent and easy to use was also an essential feature of the project. The network structure of the institutional set-up was seen to be important, as it allowed the teams to interact, and to share information and research findings.

The SARP methodology was based on multidisciplinary interaction and collaborative research between agricultural research institutions in LDCs and developed countries. It was intended that the benefits of using crop simulation models would become evident to scientists and management staff at NARCs, through collaborative action on particular problems. In 1992, this objective was partially fulfilled, as 13 out of 15 teams were still using crop simulation models in their own research, 8 years after the first training session was undertaken in 1986 (ten Berge *et al.*, 1994).

The SARP project had impact in terms of both skill development and scientific output (Mutsaers and Wang, 1999). The participants of the three training periods numbered about 90 people from 16 NARCs in eight different countries (India, Bangladesh, China, Korea, Thailand, Malaysia, Indonesia and the Philippines; ten Berge *et al.*, 1994). Among the participating centres were a number of universities, some of which now offer basic courses using the systems analysis approach in their programmes. More than 300 publications in one form or another resulted from the project (R. Rabbinge, Wageningen Agricultural University, 2001, personal communication), some of which appeared in a special issue of Field Crops Research (volume 51). These covered areas such as germplasm evaluation, input optimization, rice–pest interactions, cropping systems optimization, and regional analysis. Work carried out by some of the teams in the project on the effects of climate change on rice production (Matthews *et al.*, 1995a) is discussed in more detail in Chapter 7 of this book. Moreover, the SYSNET project, described in Chapter 6, which aimed at developing and evaluating methodologies and tools for land-use analysis (Roetter *et al.*, 2000a), was a follow-up project involving previous SARP teams from the Philippines, Malaysia and India.

However, in a subsequent evaluation of the project by Mutsaers and Wang (1999), a number of limitations were identified which have relevance for future such projects:

1. *Lack of model-building capability in NARCs*: The transfer of modelling skills to collaborating scientists in NARCs was a major component of the SARP project. Although most of these scientists were able to use crop simulation models, their ability to contribute to their development was limited. Model-building skills need to be supported for long periods of time, if they are not to be lost altogether.

2. *Decline of model use in NARCs*: A decline in the use of models in research was a problem. While the most skilful NARC scientists continued to use the models in their research, it was essential for them to be able to continue interacting with scientists at advanced institutions. The least skilful members were in danger of losing their modelling skills altogether without continuous input. The need was therefore evident for advanced institutions (either within developed or developing countries) to supply these skills, so that NARCs could continue to use crop simulation models in their research.

3. *Lack of impact of crop simulation models on applied research in NARCs*: Although the SARP project did have impact on scientific output and skills development in terms of the numbers of publications as described above, the expectation that models would significantly increase the efficiency of applied research in NARCs still appears to be 'more promise than substance'.

4. *Inability to simulate complex cropping patterns in less developed countries*: Process-based models cannot always capture the full complexity of farming systems in LDCs, and may adequately handle the effect of only a few variables, in uniform monocrop systems. This has made them more useful in developed countries, where conditions can be more easily controlled. The relevance of such models to resource-poor farmers may therefore be limited.

5. *Lack of relevant focus*: The concerns driving some of the models, such as maximizing yields, are not always relevant to farmers whose objectives may be different, for example, maintaining the stability of yields.

The SARP project foresaw the need for comprehensive education and training of NARC scientists as part of the development of modelling skills and use of crop models in general. It planned carefully for a devolving process that would result in an independent network of multidisciplinary teams at various NARCs working with IRRI. However, despite the scale of the project and the foresight that appears to have gone into it, Mutsaers and Wang (1999) found that modelling skills were being lost, and that the use of models is not likely to continue unless there are continued interventions from 'advanced' organizations. Several general questions inevitably arise, therefore, in connection with the observations made by Mutsaers and Wang (1999), that are relevant to the use of any crop–soil simulation model in developing countries:

1. How long can support from 'advanced' organizations be maintained, given that the SARP philosophy was to create self-sustaining model-building and model-use capability in NARCs?
2. Are the modelling tools too complex to be taught to scientists whose primary aims may not be modelling? Would an alternative approach be better?
3. Can NARC scientists make their research relevant to resource-poor farmers rather than perpetuating a research tradition that may be more relevant in orientation and method to agriculture in developed countries, or to richer farmers in developing countries who can control their cropping environment? Can modelling better simulate the objectives and perceptions of resource-poor farmers?
4. Can modelling capture the complexity of farming systems with multidimensional spatial and temporal variability?
5. What is the opportunity cost of developing the skills required to build and to use crop simulation models in NARCs? Would the benefits for poor farmers be greater if the money was spent differently?

12.4.3 The REPOSA project

The REPOSA (REsearch Programme On Sustainability in Agriculture) project was a 12-year cooperative programme from 1987 to 1999 between Wageningen Agricultural University (WAU), the Tropical Agricultural Research and Higher Education Centre (CATIE) and the Ministry of Agriculture and Livestock (MAG) in Costa Rica. Some of the work carried out under this project has already been described in Chapter 6. A full description of the project and its overall achievements, along with a frank analysis of the lessons learned from the experience, is given by Jansen (2001), which we summarize here.

The project originated from a desire by WAU to establish a base in the humid tropics to carry out both research and education by Dutch scientists and students, and as such was largely supply driven. Two phases were distinguished – in the first, from 1987 to 1991, the research was mostly monodisciplinary (i.e. soils, agronomy, sociology, economics), whereas in the second, from 1992 to 1998, a much more interdisciplinary approach was taken. This change in focus arose from the realization by research managers at WAU that in order to answer the types of questions that society was asking of agricultural researchers, a single disciplinary approach was not adequate. The project, therefore, focused on the integration of biophysical and economic disciplines, taking into account spatial issues, which led to the development of a general framework for land-use studies on the subregional level. This framework was called SOLUS (Sustainable Options for Land Use) (see Chapter 6, Section 6.1).

Outputs of the project in the first phase consisted mainly of traditional soil science research publications and an extensive GIS database, although no new scientific concepts were developed. The second phase, however, saw the development of new concepts as part of the interdisciplinary approach to land-use analysis. For example, soil scientists developed and applied the concept of functional soil layers that enhanced the speed and efficiency with which digital soil maps could be produced. Innovative approaches to linking data from different scale levels were also developed. Some 350 publications of various kinds were produced, including 70 refereed journal articles, about 100 MSc theses and a number of PhD theses (Jansen, 2001). Over the duration of the project, about 235 students were involved in its activities, including students from WAU and other Dutch higher education institutions, CATIE, and similar institutions in Costa Rica. Although capacity building was not an explicitly formulated goal, a number of activities towards this end were carried out by REPOSA staff, particularly near the end of the project. This consisted mainly of supervision of students on CATIE's MSc programme and *ad hoc* lectures by REPOSA staff to existing graduate courses at CATIE.

Despite this, engagement with stakeholders seems to have been less than hoped for, for a number of reasons (Jansen, 2001). These included land-use analysis not being a research priority of CATIE's, a divergence of research paradigms between WAU and CATIE (i.e. monodisciplinary versus interdisciplinary), and limited involvement of REPOSA staff in CATIE's educational programmes. Cooperation between REPOSA and the Ministry Agriculture and Livestock was also limited due to the latter's lack of human and financial resources, and its mandate for rather more applied research. Nevertheless, one MAG officer was registered for a PhD at WAU, and his research produced useful practical results. Also, at the end of the project, the SOLUS methodology was applied to a pilot watershed in the Central Pacific Region of Costa Rica to demonstrate it as an instrument for policy support and an aid to discussion amongst stakeholders with contrasting objectives. Although interest in SOLUS by top Ministry officials had been minimal up to that point, they viewed the results of this exercise sufficiently positively to develop a follow-up project for training in the SOLUS methodology. However, Jansen (2001) makes the point that methodologies for land-use analysis take some time for development before end-users can become involved, and that this late interest is perhaps to be expected.

Jansen (2001) summarizes the lessons learned from the REPOSA project as follows:

- Tools for land-use analysis should have a multidisciplinary and multi-level character.
- Effective use of these tools will be enhanced by regular and close inter-action among stakeholders in general, and should be right from the start, in order to develop their full support. However, the methodologies need to be in an advanced stage of development for them to be actively involved in their use. How to reconcile these two opposing conclusions needs to be thought about.
- A substantial investment in training is necessary to ensure that potential users gain sufficient understanding of the methodologies.
- Intensive consultation with policy makers and other stakeholders is nec-essary for the identification and evaluation of relevant policy scenarios to which to apply the methodology.

12.5 Considerations for Education and Training

In many ways, considerations for education and training can encompass issues that are relevant to both research and decision making, as students may go on to become either researchers, decision makers or educators. At the same time, however, there are those issues which are specific to education and training, in particular those concerned with providing information that is relevant and stimulating with the appropriate teaching tools and methods.

When crop simulation models are used as tools in education and training, they are usually only as good as the tutors who use them. Learning through CAL tools can be frustrating and sterile for students, when tutors dissociate themselves from the teaching process altogether, in the expectation that the model will replace them. Additionally, Pilkington and Parker-Jones (1996) have suggested that if students are to make significant cognitive advances, it is important for tutors to be present, acting as role models by externalizing solutions to particular problems; these then guide the future thinking of the students.

Clearly, what is important in education and training is the ability of the crop–soil simulation model to convey a message to students, and the correct use of the model as an educational tool by the tutor. Models to be used for teaching and training have different requirements from those developed for research or decision support. These can be divided into considerations for the underlying mathematical model, and those relating to the packaging of the software.

12.5.1 Model considerations

Crop simulation models can be divided into two major groups, the 'scientific' or mechanistic and the 'engineering' or functional groups (Passioura, 1996). This distinction is extremely important as it can determine which model should be used in a particular context. Mechanistic, process-based models are typically used in scientific research to improve understanding of crop physiology or crop–environment interactions. Functional models, on the other hand, are typically used in an engineering context, where solutions are required for practical problems. Given the distinction that Passioura (1996) makes between mechanistic and functional models, it is clear that there are implications for the educational process.

Mechanistic models are increasingly used in the educational process to replace real crops. However, Passioura (1996) is dubious about the unconsidered use of such models in this way. The mechanistic model, for example, is often unsuitable for practical problems that require practical solutions. It follows then that students should not be taught to use such models without due consideration of the type of the problem being tackled. For example, for a specific location, it is probable that a simple, empirically derived, functional crop simulation model, possibly based on allometric relationships, will give better results.

During the early adolescent phase of computer modelling (see Chapter 1), when optimism about the potential of crop modelling was high, the belief that universal models could be developed, explaining crop growth anywhere at any time, was thought by many to be possible. This led to complex, reductionist, mechanistic, process-based models. Sinclair and Seligman (1996), however, point out that it is important to dispel the myth that computer models can accurately simulate crop growth everywhere in time and space. Due partly to our incomplete understanding of natural processes, but also because of the technical difficulty of fully modelling natural resource systems, it is very difficult to develop a single model, capable of simulating crop growth universally, without calibration adjustments or further modelling.

Mechanistic models are often complex, and errors within them are compounded to the extent that they must be calibrated to specific systems to achieve accurate results. This is problematic in that it ignores the obvious fact that the model does not explain the growth of the crop, and may therefore be erroneous in structure or process. Additionally, mechanistic models are often based on a multitude of hypotheses and untestable 'guesstimates' (Sinclair and Seligman, 1996). In this sense, they violate the scientific process (hypothesis – experimentation – theory), which requires that a hypothesis is subjected to practical experimentation and measurement in the real world. As Passioura (1996) points out, 'Good science involves developing a set of views about the world that are subjected to penetrating experimental (or observational) testing.'

Finally, there is also the danger that students taught entirely through crop simulation models, and who are not taught to critically evaluate the structure and processes within the simulation, may end up by believing that the model represents some kind of real world 'truth'. As Passioura (1996) suggests, 'unless they have been exposed to real as well as simulated plants, . . . fantasies in FORTRAN can too easily become fact'. However, he does indicate that there are important educational benefits to be derived from mechanistic models, precisely because they cannot accurately simulate every situation. The explanation for these discrepancies can often lead to conceptual breakthroughs in scientific thinking, which are educational for the scientist in a research context, but may also then feed back to students through journals and text books.

Passioura (1996) suggests that functional crop simulation models may be more useful as practical tools than mechanistic models, because they are based on robust empirical relationships between plant behaviour and major environmental variables. Sinclair and Seligman (1996) have also pointed out that some of the most accurate predictions of crop growth are made by simple models, or even empirical equations. Given our lack of understanding on many of the processes occurring in crop growth and crop interactions with the environment, Passioura (1996) suggests that it may be futile to use mechanistic models to help make decisions, for example, for farm management. Functional crop simulation models would probably be more satisfactory in such 'engineering' or problem-solving contexts. Additionally, functional models can be more robust, are simpler, require smaller data sets and are easier to use and understand than mechanistic models.

The classification of crop simulation models as either mechanistic or functional has implications for education and training. The appropriate choice of model will depend on the educational context. Thus a mechanistic model might be usefully employed to teach an in-depth understanding of plant nutrient assimilation. It might, for example, be used by students training to be future researchers, as for them an in-depth understanding of the physiological processes embodied by the models would be appropriate. A mechanistic model might also be usefully transferred to researchers at research institutes in LDCs, to further their own scientific learning and research. A functional crop simulation model on the other hand, being largely empirical, or even based on allometric relationships, would be far better used by students who will have to make practical decisions or manage agricultural systems in the future. This could include students being trained for decision making or agricultural extension.

It is clearly important for anyone using crop simulation models in education to be aware of the importance of the distinction between mechanistic and functional models. Indeed, the whole debate surrounding the use of crop simulation models is an important one for tutors and students. Where crop simulation models are used as tools in education and training, students should, as a matter of course, be familiar with these issues. For example, functional models should generally not be used without further development, outside their intended target area, as they are very site-specific. They have to be modified or calibrated to suit the conditions of each new context. Mechanistic models might be better suited to simulating plant growth over a wider range of conditions, although the 'accuracy' of the results may be less than that obtained from a functional model specifically developed for the location. A student should also know that in extreme environmental situations, a mechanistic model might have to be redeveloped or calibrated to produce useful results.

Further considerations may extend to providing students with an understanding of the more philosophical and wide-ranging issues raised by the development of simulation models in general. How far have they been able to help as decision-making tools? To what extent have they proven to be useful as representations of reality? To what extent should we believe what they tell us? To what extent can we trust the internal computer algorithms and the conceptual structures of the models? Might they be giving us the 'right' answer for the wrong reason? When should mechanistic and when should functional models be used? Is the real value of models in helping us to bring rigour and form to areas that are hazy and unclear, rather than in the results that we can generate from them? Many interesting papers exist in this area (for example, Philip, 1991; Baker, 1996; Boote *et al.*, 1996; Monteith, 1996; Passioura, 1996; Sinclair and Seligman, 1996), and students would benefit greatly from being familiar with these issues – not simply because such questions are interesting in themselves, but also because they will help them in future to treat models with an appropriate degree of caution and scepticism.

To some extent, it should be recognized that simulation models for teaching and learning purposes have different objectives to those used in research or decision support. The emphasis should be on the learning outcomes rather than the accuracy of the model predictions. So long as the model is based on a correct understanding of the mechanisms involved, and so long as it responds in the right 'direction' and with the appropriate order of magnitude, producing relative values that are acceptable, then absolute accuracy may not be essential.

12.5.2 Packaging considerations

The issue of how transparent or opaque a model is, is important if crop–soil simulation models are to be used as tools in education and training. Transparent models allow students to examine their structure and processes; this allows them to understand the dynamic interactions of its internal logic and facilitates the process of learning (Sinclair and Seligman, 1996). For example, to understand cause and effect, the student may need to be able to 'look inside' the model to see why things are happening in a particular way. Opaque 'black-box models' that merely present a user interface are not subject to the depth of scrutiny that may be required for effective learning outcomes. The assumptions and logic of such models may be difficult to determine and much of the teaching impact is therefore reduced.

However, transparency may not always be needed and if the objective of the learning exercise is merely for the student to develop a 'feel' for the impact of inputs on an output, then a black-box model may be quite acceptable, especially if this affords advantages in terms of speed of execution or parsimony. By observing the change in output for a given change in input, such models allow the student to develop simple heuristics that can be applied in other areas of their studies. For example, allowing students to observe how crop yields change as levels of water and fertilizer are changed should lead to an understanding of the concept of limiting resources and water–nutrient interaction, without requiring a full understanding of the science involved in water and nutrient dynamics.

Students should find using the software clear, simple, intuitive and flexible. It should require the minimum amount of time to learn how to operate it, so that the majority of their learning time can be devoted to the twin processes of 'virtual' observation and experimentation. There have been many examples where models developed for other purposes have been used in education and students have struggled with the interface. For example, Abraham Singels (S.A. Sugar Association, 2000, personal communication) commented, 'we used the PUTU suite of models for practical training of students in the crop-modelling course. Students generally struggled to master all the intricacies of the menus etc. in the short time available.' In such circumstances, students can then mistakenly perceive that the objective of the exercise is simply to get the model to work! The deeper lessons relating to the understanding of the system's behaviour can become obscured.

The use of a standardized interface, such as Windows, should make the software more intuitive to the student who may already be familiar with word processors or spreadsheets. There is now a generation of graphical 'shells' being developed for old (but useful) FORTRAN models that handle the input and output in a user-friendly manner, but retain the integrity of the original code.

Inappropriate style, including the language of the interface, can be a great distraction to the student. For example, the DSSAT shell has been used to teach undergraduate students in Thailand. Simulation modelling was perceived to be an integrated and practical approach to teaching and feedback from the students was good. However, difficulties stemmed from the documentation and programmes being in English, rather than in Thai (A. Jintrawet, University of Chiang Mai, Thailand, 2000, personal communication).

Many potentially suitable packages are off-putting due to the system of units and symbols being used. Courseware developed for the American market, using Imperial units, is unlikely to be accepted in the European market, which is based on the SI system. Even between disciplines, preferences for units differ. This can be overcome, as for example, in the SPACTEACH package (see Section 12.2.2), which gives the user the choice of working with soil water potentials in cm, J kg^{-1} or kPa. All input and output will be in the chosen units. Any language, symbol or unit that is unfamiliar to the student will make it more difficult to achieve the underlying learning objective.

Where models have been used 'off the shelf' there is a danger that the tutor is not fully aware of the limitations and assumptions of the model – this can be due to poor documentation. Students, when allowed to experiment freely, may push the model beyond its intended range and start to produce inappropriate results. A well-produced package would alert the user to this and constrain his/her input values, with error- and range-checking functions.

Speed of program execution is an important characteristic of models to be used as teaching and learning tools. Whereas a researcher may have the time to wait for long model runs to complete, a student may not. If they cannot see the results within a few seconds, or at most within a few minutes, their attention, and the message, may be lost.

In summary, important features for educational use include some of the following:

- intuitive graphical interface;
- dynamic animation of the processes being simulated;
- graphical representation of results;
- the ability to pause a simulation and interrogate the state of variables within the model;
- the ability to select appropriate units;
- speed of execution;
- error and range checking of input data;
- default values for parameters and hints and tips on how to derive their values;
- on-line help that explains the science behind the model, not just the mechanics of operating the software;
- the ability to export results, tables and graphs for reports and statistical analysis packages.

12.6 Limitations and Constraints of Models in the Educational Context

The limitations or negative impacts of using crop simulation models in education and training should not be ignored. Furthermore, there are constraints that can make the integration of models in institutions for education and training difficult. These may combine to make the use of crop simulation models problematic and even undesirable, and it is for institutions and tutors to decide whether the use of crop simulation models as tools for education and training suit their particular circumstances.

12.6.1 Inappropriate substitution of the real world

By using crop simulation models, students are separated from the real-world phenomenon that they are supposed to be studying and this may lead them to believe that the model is 'reality' (Philip, 1991; Passioura, 1996). However, models are simplifications of reality, and if incorrect or inappropriate assumptions have been made in the model, misleading results can be produced. McLaren and Craigon (1981) noted that for TRITIGRO I, that as the model's output was based on 'a combination of published responses, experimental facts, and practical experience', some of the responses of the model were inevitably 'subjective' and limited by the 'ideology and ability of the authors'. Both students and tutors need to be aware of the fact that computer models are only simulations of reality, and should not be confused with reality itself.

12.6.2 Limited availability of simulation models and documentation

With traditional paper-based approaches, the tutor may often refer the student to a range of textbooks, with different approaches and styles. However, the development of bespoke packages for in-house teaching is expensive and the more cost-effective solution is to use 'off-the-shelf' crop–soil simulation models. Universities and researchers may be unwilling to invest in the development of teaching tools in the same way that they would invest in a product aimed at a more commercial market.

The range of crop–soil simulation models suitable for educational purposes is therefore likely to be limited. Much of the educational software has been developed by 'programming amateurs' and lacks a professional interface. The models that are available may not fit the local curricula or suit the teaching approach of the tutor. Furthermore, the degree of complexity or simplicity of the available models may not be compatible with the curriculum of the course.

12.6.3 Loss of field and laboratory skills

Students using simulation models have often expressed concern at the loss of real-life hands-on experience of the 'tools of the trade' for measurement and recording (McAteer *et al.*, 1996). While the use of models may be an efficient use of the student's and the tutor's time, at some stage students still need to learn the practical skills. Philip (1991) noted that the loss of field and laboratory skills corresponded to the development of modelling as a substitute for experimental and field investigation. Models should therefore be used in conjunction with traditional field and laboratory work, so that essential field and laboratory skills are not lost.

12.6.4 Cost effectiveness

Good simulation models, suitable for educational purposes, are expensive to produce and the educational market is relatively small (Thomas and Neilson, 1995). The devotion of resources to software development is often not perceived to be cost-effective. While computer simulations may appear to be a cost-effective alternative to laboratory equipment, 'its real monetary cost is often concealed by accounting systems friendly to computing', and 'it is the concealed non-monetary cost which is troubling' (Philip, 1991). Evidently, there are significant support costs for simulation modelling, beyond the simple hardware and software costs.

The costs of tutor time and the production of supporting materials is often overlooked. The level of tutor support required is often underestimated (see Section 12.6.5) and anticipated savings in staff time (and cost) often fail to materialize. The software needs to be supported by other teaching materials. In Edward's (1996) example, the computer-based simulation was packaged with a workbook, a video of the real system being modelled and a results sheet. These materials may have to be produced locally to ensure relevance and compatibility within the curriculum of the course, and can take much time to develop.

12.6.5 Lack of proper tutor support

Substantial tutor support is still required for effective learning to occur through the use of crop simulation models. The best use of computer simulations in education and training appears to be where the models are properly integrated into the teaching programme (Edward, 1996; McAteer and Craigon, 1996) and supported by introductory lectures and plenary sessions (CLUES, 1996a). For students, learning through computer simulations can be a frustrating and sterile experience where tutors dissociate themselves from the learning process, in the mistaken belief that the model will do all the teaching for them. Edward (1996) concluded that the effectiveness of simulated labs was greatly enhanced by the active engagement of the tutor. Thomas and Neilson (1995) suggested that the tutor needs to be present to guide the students, prevent wasting of time, help in critical evaluation and prompt the formation and testing of hypotheses.

12.6.6 Lack of cognitive development

Pilkington and Parker-Jones (1996) have commented that a danger inherent to the use of computer simulations is in the 'setting of tasks which can be completed by attention to surface characteristics without making higher level cognitive demands'. This can result in the tendency for students to manipulate screen objects without developing deeper insight into the principles that underlie the model's observed behaviour. The solution is to lead the student into cognitive conflict by setting them tasks that are difficult to solve within their current knowledge. This forces students to 'make significant changes to the way that they interpret the world'. However, this in itself is insufficient, as students must recognize and resolve such cognitive crises. The role of dialogue with peers and, more particularly, with tutors is vital, as this allows students to follow a role model who externalizes 'modes of thinking and reasoning which learners come to internalise and use to guide their own thinking'.

12.6.7 The 'not invented here' syndrome

Educational software developed at one institution has often failed in transfer to other sites (Thomas and Neilson, 1995). Teaching staff at one institution are reluctant to use courseware that has been developed elsewhere. Often this may be because the software has not been well documented and the tutors may not feel that they have sufficient understanding of the internal workings of the model to: (i) verify its relevance and (ii) explain to students why particular results are generated. Acceptance is more likely if models are fully and explicitly documented and evaluated by an independent third party.

12.6.8 Constraints in the educational system

Despite a lot of talk about CAL and IT generally being the future of education, even in the 'developed world' their use is far less widespread than might have been anticipated. Loveless (1996) for example, found that in US public schools, computers were not as widely used as expected. This had often been mistakenly blamed on 'recalcitrant bureaucracies and stubborn teachers'. However, they found that the true cause was more frequently rooted in organizational constraints of the school system and the nature of teachers' and students' work. Similarly in the UK, Scott and Robinson (1996) concluded that the use of IT in education involves not just a change of teaching resources but also of teaching strategies and beliefs.

The successful adoption of IT by schools in the UK was directed by the Education Reform Act of 1988 but was part of a broader change in the educational system as a whole brought about by that Act. Thus the adoption of IT was enforced by the national educational policy and facilitated by funding for training and resources. In contrast, there appears to have been little evidence of such a change at higher education level, despite substantial government funding for initiatives to develop computer-based teaching and learning at university level.

The uptake of CAL for education and training has had a difficult teething period even in the UK. The difficulties are likely to be far greater in educational institutions in LDCs and there may be little to be gained by trying to develop a method of teaching that simply cannot be supported by the host organization. The development of institutional capability in LDCs may therefore be more important in the immediate future than the attempt to use crop simulation models, as tools for education and training.

The challenge for developing countries is to ensure that donor perspectives do not determine the agenda for their education and training organizations. The specific needs of LDCs should not be ignored, or worse still, assumed to be the same as those of developed countries. Crop–soil simulation models are most likely to be successful tools in education and training, if the needs of LDCs are properly understood, and the often complex issues raised by their use are properly dealt with.

As a general rule, the transferral of technology without the appropriate integration of that technology into the local context, has rarely proved to be successful. The use of crop–soil simulation models in education and training in LDCs needs therefore to be properly contextualized and managed, as transferral is not in itself sufficient. The successful use of CAL in developing countries will not only require the provision of hardware and suitably designed courseware, but the development of supporting documentation and a change in the approach to teaching and learning in LDC educational systems.

Clearly, there may be a high opportunity cost in attempting to integrate the use of crop–soil simulation models in education and training and the net benefits need to be carefully weighed to determine if they are worthwhile. There may be more effective ways of improving the quality of education delivered to students, than through the introduction and use of crop–soil simulation models. These alternatives, as well as the *status quo*, should always be considered.

Part 4
Have crop models been useful?

Who are Models Targetted at? 13

Robin Matthews

Institute of Water and Environment, Cranfield University, Silsoe, Bedfordshire MK45 4DT, UK

In the preceding chapters, we have reviewed a number of applications of crop–soil models in tropical agriculture, ranging from purely research applications, applications that were designed to aid decision makers, through to applications of models in education and training. In all of these, there were target groups whom the model developers may have had in mind when they constructed their models. In this chapter we discuss in more detail the groups of people who might be expected to benefit from the use of crop–soil simulation models.

In Fig. 13.1, we have attempted to show in simplified form the relationships between the main stakeholders in the context of developing, using and benefiting from the use of models of agricultural systems in developing countries. The linkages between the different groups can be thought of as flows of information – the solid arrows represent the flow of information contained within a model itself, whereas the dotted arrows represent the flow of information that has arisen from the use of a model, but is in a form other than within the model, such as a report, pamphlet, poster, verbal communication (e.g. personal contact, radio, TV, etc.) and so on.

The first level represents the people involved in developing the models, which generally seem to be scientists in developed countries, or scientists working at the International Agricultural Research Centres (IARCs). The second level represents the direct users of models, i.e. those who actually take a model and run it and interpret the results. In our classification, these may be consultants (in either developed or developing countries), scientists in developing countries and educationalists in developing and developed countries. The third level represents the groups of people who may

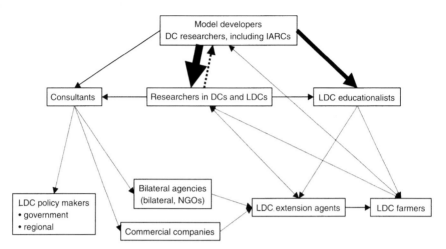

Fig. 13.1. Postulated relationship of crop simulation models to flows of information between various stakeholders in agricultural development. Solid arrows represent information encapsulated in models themselves, dotted arrows represent flows of information by other means, but arising from the use of models. Thickness of arrow is an estimate of relative information flow rate. Overlap may occur in cases, e.g. researchers may also be consultants. DC, developed country; LDC, less-developed country; IARC, International Agricultural Research Centre; NGO, non-governmental organization.

potentially benefit from output from the models, but who are unlikely to use the models directly themselves. We recognize that the boundaries between the groups are not clear cut, and in many cases, the same people may be fulfilling two different roles; for example, model developers may also be using their models in an educational role in graduate classes, or as tools for consultancy work. Similarly, there may well be cases where people within development agencies or non-governmental organizations (NGOs) may directly use the models. Nevertheless, as a workable framework for thinking about how models may fit into the overall development process, we believe it is a useful starting point.

In the remainder of this chapter, we discuss each of these groups in more detail.

13.1 Researchers

There is no doubt that, so far, the largest uptake and use of models has been by the research community, both in developed and developing countries, although this does seem to be skewed towards the former. This is because models are primarily research tools – for most scientists, the incorporation of knowledge into simulation tools is a process taken for granted. For them, using models is a way of organizing and utilizing information that seems obvious.

However, within the research community, there are many disciplines, and it is perhaps a little misleading to talk about them as one group. For example, Stephens and Hess (1996) list the participants in a crop modelling workshop as including agronomists, plant breeders, agrometeorologists, agricultural economists, soil scientists and specialists in soil conservation, crop protection and irrigation, as well as university lecturers and post-graduate students. All of these may potentially benefit from the use of crop–soil simulation models, but the needs of each will be different. Some may be interested in taking a model and adapting the source code to their own particular use, others may just want to take a model 'off the shelf' and use it without further modification.

In developing models for researchers to use, thought needs to be given to the type of questions that could be answered by the use of a model, whether they would be using the model themselves and, if not, how the information from the simulation will be transferred to them. By clarifying these issues, it will be easier to develop a system that is relevant to their use. Care also needs to be taken that the target group has the ability, willingness, and hardware to use the proposed system. It is also important to ensure that there is long-term technical support available and that the software can be responsive to immediate updates, corrections and improvements (Knight, 1997).

The impact of work done at the researcher level on addressing poverty in developing countries is more likely to be longer-term and less easily quantified. Nevertheless, it is no less important because of that. Shorter-term impact may be achieved by having a clearly defined problem, and for researchers and farmers to work more closely together. This approach has been argued for by an increasing number of authors (e.g. McCown, 2001), and is starting to be tried by various groups (e.g. the APSRU group, see Section 10.6). Farmers and researchers become partners, and collaborate to research best practice in the context of current practice, using theory and models where appropriate. This point is discussed in more detail in Chapter 14 on model impacts.

13.2 Consultants

It is difficult to know the extent to which agricultural and policy consultants use crop–soil simulation models, as generally such work is not published due to client confidentiality. Examples of the DSSAT models being used in consultancy work in South Africa has already been discussed in Chapter 11. The point was made there that the modelling work was carried out on behalf of the clients by the modelling consultancy – the clients did not use the models themselves.

Again, there is probably a range of needs of model characteristics, depending on the type of consultancy work being undertaken. Most consultants may just wish to use the model without further modification, in which case they will want a user-friendly interface with the ability to enter the options they are interested in using the model to explore, and to obtain the output that they want in an easy-to-understand format. In the case where scientists with modelling ability are also carrying out consultancy work, a user interface maybe of much less importance, and the ability to 'get inside' and modify the model to their own requirements may be an essential requirement.

The impact of the work done by consultants on addressing poverty issues in developing countries is more difficult to evaluate, and depends on the nature of the work carried out. If it is for a private company, it is less likely to help poor farmers, although by helping a company to improve its efficiency by making better decisions, local employment may be enhanced. An example of this is the tea industry in Kenya, Tanzania and Malawi. If the consultancy work is carried out for an NGO, the impact on poverty could be more immediate and direct. If it was for an international aid or development organization such as the Department for International Development (DFID) or the United States Agency for International Development (USAID), the potential impact of making well-informed decisions could be large, as in the example of fertilizer to Albania (see Section 9.5).

13.3 Educators and Trainers

Educators and trainers are other important end-users of crop–soil models. As discussed in Chapter 12, they may either use models to help illustrate to their students particular crop physiological processes (e.g. photosynthesis, water uptake, etc.) and the effects of these at higher levels, or they may be part of the process of learning systems analysis techniques, in which case the students may build their own models or use existing models to provide information about a component of a larger system, such as a farm or a region. The needs in each case will be different – if it is as a teaching aid to illustrate particular processes, it is important that the model has a user-friendly interface that can present the information in a way that the students can easily understand. Absolute accuracy may not be so important as the ability of the model to represent the shape of the particular response curve realistically. Where students are learning the modelling process, a model that allows them to concentrate on the processes being modelled rather than being concerned about programming issues may be an important consideration. Other requirements are discussed in Chapter 12.

The impact of education and training on poverty issues is, again, likely to be long-term, but education is an essential investment if developing countries are ever to develop sustainable means of producing their food. Learning to think in a 'systems' way rather than just focusing on a small specialized area is necessary if any impact of research and education is to be made. The use of models and training in systems-analysis techniques such as in the Simulation and Systems for Rice Production (SARP) group (see Section 12.4.2) must contribute to this.

13.4 Policy Makers

There may also be scope for the use of simulation models to support strategic decisions at a larger scale. However, the extent to which this is successful will depend on institutional issues and the level of training available to the decision-making staff. In developed countries, there tends to be a greater commitment to acceptance of new knowledge and promoting new practices. This allows technology to be advanced more rapidly (Tollefson, 1996). In the institutions of less-developed countries (LDCs) there may be resistance to new technology especially if it is seen to pose a threat to the existing system. Scott Swinton (Michigan State University, 2000, personal communication) suggests that if decision makers are very busy and already dubious about the value of models, they may find the opportunity cost of learning unacceptable.

Many agricultural policy issues do not require the use of models and agroecological information, and to pretend otherwise could result in a wasteful use of research funds. For example, reforming agricultural banks, strategies for privatization of state-owned agricultural facilities (e.g. grain silos, mills, etc.) and ways of restructuring the operational basis of producer cooperatives can all be dealt with without models or agroecological data. However, there are potential areas in which models can make a contribution to policy making. Norton (1995) has proposed the relationship between crop research, agroecological research and agropolicy research shown in Fig. 13.2. He suggests that we need to think about exploring the intersection between agroecological research and agropolicy research, i.e. the set where agroecological knowledge is relevant to the formation of agroeconomic policy. It should be remembered that the interest of policy makers is not so much with the best state of an agroecological system, but with the effect of the type of interventions they have in mind. Possible areas of intersection are:

- *Determining comparative advantage* – lack of this in particular crops will depress the incomes of those who grow them – will lead to shifts in cropping patterns and/or increased rural/urban migration.
- *Making predictions of producer responses to policy changes* – e.g. incentives, resource endowments, access issues.
- *Helping design policies for sustainability* – prediction of future changes in the natural resource base.

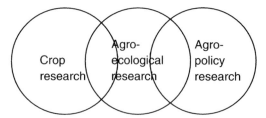

Fig. 13.2. The relation between crop research, agroecological research and agropolicy research (from Norton, 1995).

He also makes the point that decision-support models need to be available at the right time and need to be accessible to the policy maker. Aggregating farm level models into regional or sector models may improve the quality of information, but may also extend the time required for model development. If a decision model is developed in response to a specific policy question, timeliness may constrain the amount of detail in the model. Policy decisions are often needed to react to a changing environment and the quality of the decision depends on the timing. For example, after new international agreements are made (e.g. GATT, WTO), new land-use options may become available in certain countries. If a country takes too long to evaluate these options, it may find its possible niche in the world market being taken up by other countries. One possible solution is to have a 'toolbox' of well tested and validated model components (e.g. modules for soil water balances, nutrient dynamics, different crops, etc.) that can be assembled in a short time to answer a specific question.

Policy makers can probably have the largest potential impact on addressing poverty issues than any other group, purely because their decisions can affect so many people. If wrong or ill-informed decisions are made, the poverty of people at lower levels can be worsened; well-informed decisions, on the other hand, have the potential to enhance people's livelihoods and lift them out of the poverty trap they may be in. It is difficult, however, to quantify this impact, and even more difficult to compare the outcomes of different decisions, as it is generally not possible for different decisions to operate concurrently with all other factors maintained constant.

13.5 Extensionists

Extension personnel are an important link in the chain between researchers and farmers. In developed countries, they often provide farmers with a human interface to computerized decision support systems (DSSs). In farming, much advice comes from trusted advisers, and to substitute this with the use of a computer for advice may be off-putting to many farmers (Knight, 1997). For example, Blokker (1986, cited in Macadam *et al.*, 1990) found that DSSs designed for direct use by farmers (as distinct from those where interpretation is done by an extension officer) were generally not appreciated by farmers and had only marginal influence on decision making. Reasons for this have already been discussed in detail in Chapter 11.

In developing countries, however, it is less likely that extension personnel will have access to a computer, in which case computerized decision support systems will not be appropriate. Information from research is more likely to reach them in other forms such as research reports, brochures, posters, training workshops, verbal communications at field days or from informal contacts with research staff. The information encapsulated in the form of a computer model, therefore, is not likely to flow in that form further than the research and education groups, but this does not mean that information flow need stop there altogether. The research–extension–farmer relationship should be viewed as an interdependent continuum. Model output can still be used by researchers to produce useful information for extension staff and farmers; it will just be in a different form. This could be in the form of simple rules of thumb regarding planting dates, or times and amounts of fertilizer applications, for example. We have attempted to show this in Fig. 13.1 with different arrows representing flows of information in these different forms. The problem, therefore, becomes a general one relating to dissemination of all agricultural research results, not just those from modelling. Unfortunately, extension services in many developing countries are badly under-equipped in terms of staff, transport and accommodation (Tollefson, 1996), not to mention access to computing facilities. This situation may well change in the future, however, as computer technology becomes cheaper and the skills to operate them become more widespread; computerized DSSs may well become more relevant to extension staff then.

The influence of extension staff on alleviating poverty can be very significant, as they are in direct contact with the farmers being targeted. It is the efficiency and speed with which they are able to transmit information to the farmers that will partly determine whether a particular technology is likely to be adopted or not. They also have another important role to play, which is the transfer of information in the opposite direction, from farmer to researcher, so that research activities are relevant to the real problems faced by farmers, not just problems that researchers perceive farmers to have.

13.6 Farmers

It seems unlikely that, in the short to medium term, there is any potential for on-farm use of computer-based DSSs by small-holder farmers in developing countries. Firstly, the financial ability to purchase and run a computer is a long way off what most subsistence farmers would be able to achieve. Many may not even have an electricity supply available to run a computer. Secondly, the level of education needed to be able to operate one successfully is likely to limit uptake. Daniels and Chamala (1989, cited in Hamilton *et al.*, 1991) found that farmers' interest in computers was related to their level of education – those with higher levels of education were more interested, while those with less formal education preferred to go by experience. Even in developed countries, poor computer literacy of farmers has hindered the uptake of information technology (IT) systems (Hamilton *et al.*, 1991). In developing countries where rural education is of a low standard, and even the educational level of extension workers is not always high, the constraints are even greater. Developing country farmers would require a huge amount of training and support to begin to use the systems in a useful way. Thirdly, as discussed in Chapter 11, it is not at all certain that the answers to the sorts of questions that farmers are most likely to ask could be provided by operational decision support systems anyway.

Seligman (1990) believes that the next farming generation in developed countries may make more use of models due to their increased familiarity with the medium. In the case of the IWS software in the UK, farmers slowly lost the need to obtain their information from an adviser. Farmers initially paid for the service of an adviser who used an irrigation scheduling model to help them plan their irrigation. However, over time, as they realized that they could save themselves money by using the computer instead of paying someone to do it for them, a demand for the software developed. Eventually the consultancy was replaced by farmers using the system them-selves on-farm (Hess, 1990, 1996). However, in a context of poverty, this process of familiarization with the technology may occur over a much longer time scale.

A more likely route to success may be to adopt the approach pioneered by the Australian project described in Chapter 10 (Section 10.6), in which the Agricultural Production Systems Simulator (APSIM) package of models is currently being used in Zimbabwe to help farmers, policy makers, exten-sion agents and researchers improve their understanding of the trade-offs between different crop and cropland management strategies under scenarios of climatic risk (P. Grace, CIMMYT, Mexico, 2000, personal communication). Rather than focusing on a particular optimal strategy, the model is used to explore the consequences of various cropping practices suggested by extension personnel and farmers themselves, who are aware of their own labour and capital resource constraints. The modelling aspect is important, as it would not be possible to undertake such analysis either on-farm or at a research station in a reasonable time frame. Researchers found it useful as it made them more aware of the constraints faced by small-holders, and suggested new lines for research. A similar approach was suggested by Beinroth *et al.* (1998) in which a model could be used to explore trade-offs between domestic requirements, irrigation demand and downstream use of river water in Colombia, with regular discussions between stakeholders to new positions to be formulated and simulated in an iterative manner until a consensus was reached.

Clearly, by interacting directly with farmers, the potential to improve the livelihoods of those being interacted with is the greatest of all the target groups discussed above. However, the numbers of people whose livelihoods are actually improved as a result may be limited if dissemination of the technology outwards is less than adequate.

Impacts of Crop–Soil Models

14

Robin Matthews, William Stephens and Tim Hess

Institute of Water and Environment, Cranfield University, Silsoe, Bedfordshire MK45 4DT, UK

In this chapter we consider whether crop–soil simulation models have made, or can make, any impact in improving agricultural systems, particularly in developing countries. In the previous chapter, we discussed uptake pathways of the knowledge contained in, or generated by, crop–soil simulation models and the target groups that could make use of this information in some way. Crop models have generally been developed in developed countries, and have then been disseminated to researchers, educationalists and consultants in both developed and developing countries, who then may have applied them to situations and problems to do with agriculture in their own and each other's countries.

All along this route, there are factors that may prevent the models making any useful impact in improving the agricultural systems they are targeted at. As a way of analysing these factors, we have grouped them into: (i) limitations a particular model may have for a specific application due to its construction or the assumptions contained within it; (ii) constraints to its widespread uptake and use by end-users; and (iii) factors preventing it making a meaningful impact on the way people think or behave, even if there are no limitations for the purpose in mind and the uptake and use of the model are good. These are discussed in more detail in the following sections.

14.1 Limitations to Use

In the chapters in Part 1, we discussed, where appropriate, limitations to the use of models, or, in other words, factors that currently limited their usefulness for the particular application they were being used for. Although these were discussed in terms of research, most of these limitations apply whether the model is being used for research, decision support or education and training. For example, the resolution of most current models may be a limitation in their use to discriminate the subtle differences between genotypes (Chapter 3). This would be a limitation for researchers wanting to use a model for this purpose, but would also be for plant breeders wanting to make decisions on which genotypes to select, or for an educator wanting to teach students about genotype characteristics. Similarly, the inability of many models to describe isolated plants or spatially variable plant densities would be a limitation to their use in investigating the effects of planting densities in farmers' fields where this was common (Section 4.3). Mutsaers and Wang (1999) note that most crop simulation models were originally developed for monocrops grown under highly uniform conditions caused by high levels of external inputs, and are not well equipped to deal with cropping systems in developing countries, whose characteristics include:

- limited control by the farmer over production factors (including water);
- management practices often aimed at risk reduction rather than yield maximization;
- limited or no use of external inputs;
- high within-farm and within-field variability;
- high weed and pest infestation;
- intercropping practices.

It needs to be remembered, however, that limitations of a model do not necessarily prevent it from being used for a particular purpose, provided that the limitations are taken into account in interpreting its output.

14.2 Constraints to the Uptake of Crop Models

At the next step on the route to impact, we need to have uptake and use of the models by different target groups. There seems little doubt that crop models are increasingly being used as tools in research in developed countries, as can be seen from the number of publications in refereed journals every year. However, their use in developing countries has lagged behind. Here, we broaden the discussion to examine some of the wider constraints to their uptake in developing country institutions.

Stephens and Hess (1996), in evaluating the uptake of the PARCH model in research institutions in Kenya, Malawi, Zimbabwe and Botswana, identified a number of constraints to its uptake, which they classified as intellectual, technical and operational. These are summarized in Box 14.1.

Box 14.1. Constraints to uptake of crop models (from Stephens and Hess, 1996).

Intellectual constraints
- No relevant application
- Not convinced of applicability
- Not convinced of credibility

Technical constraints
- Haven't got the disk
- No access to computer
- Couldn't understand the program

Operational constraints
- Couldn't obtain meteorological data
- Couldn't calibrate locally used cultivars
- Lack of technical support
- Lack of intellectual support
- Didn't believe results

The intellectual constraints were common across all of those surveyed. In some cases, these arose from not understanding the relevance of models as research tools. For example, it was commonly believed that for a model to be of any value, it should be able to simulate what actually happens in a farmer's field. The relevance of a model as a means of investigating complex interactions, or extrapolating the results from field trials, had often not been considered. Apart from a few cases where the disks containing the model were not accessible, few technical constraints to the use of the PARCH model were found. Lack of suitable computer hardware was only rarely a problem, and most of the organizations were very well equipped with computers. In terms of operational constraints, some of the users that were surveyed said they often felt daunted by the breadth of data, and in the case of the cultivar parameters, the amount of detail required. Scientists frequently felt that they did not have access to meteorological data in the correct format. Many said that they did not know where to go to get technical and intellectual support and to share experiences.

Similar constraints to uptake were voiced by less-developed countries (LDC) researchers during a general discussion at a conference in the Hague in 1993 (Goldsworthy and Penning de Vries, 1994: 249), who additionally mentioned: (i) lack of validation of the models for local conditions (it was felt that this should be done by the model developers rather than the users!); (ii) difficulties in parameterizing the models for local conditions; (iii) the models were sometimes unnecessarily complex; (iv) documentation was often poor; and (v) the models did not provide answers to the questions being asked.

Most of these constraints could be addressed by adequate support to those using the models. Lessons can perhaps be learned from the International Benchmark Sites Network for Agrotechnology Transfer (IBSNAT) experience (see Chapter 12) and the Decision Support System for Agrotechnology Transfer (DSSAT) family of models, which are probably the most widely used crop simulation models in the world today. Part of this success was no doubt due to the size of the project, and the participatory and interactive relationship between model developers and model users during its course, but the continued uptake and use of the models must be due to the support still available, even though the project ceased in 1994. Users and developers still keep in touch via a list-server, so that there is a broad base of support not dependent on one or two people. Users with a problem can post a query on the list-server, and usually within a day can receive help and advice from other users or the model developers.

Overall, therefore, we would argue that uptake of models by researchers, and to a lesser extent by decision makers and educators, has been good, as indicated by the large number of model applications described in earlier chapters of this book. Closer examination, however, shows that this uptake is heavily biased towards users in developed countries. A quick scan through the affiliations of authors of papers in a recent volume of *Systems Approaches for Sustainable Agricultural Development* (Bouma *et al.*, 1995), for example, gave 58% affiliated to institutions in developed countries, 20% to International Agricultural Research Centres (IARCs) and 22% to national institutions in developing countries. While not too much should be read into these figures, it does illustrate where much of the research effort using modelling approaches is coming from. To what extent this is due to the larger number of potential users in developed countries compared to developing countries is not certain – it would be interesting to compare the numbers using models in developing countries with those in developed countries, expressed as a proportion of the total number of potential users in each.

This is not to say that the constraints to uptake just discussed are unimportant; indeed, it is clear that more effort needs to go into addressing them if the numbers of model users in both developing and developed countries are to increase, which must happen if there is to be any impact. For some crop models, the constraints to uptake are greater than others – the PARCH suite of models, for example, has enjoyed much less success in uptake than, say, the DSSAT family of models has, reasons for which have been discussed earlier.

14.3 Characteristics for Impact

The question arises, however, as to how much impact has been achieved by this uptake, particularly in relation to tropical agriculture. First, we must define what we mean by impact. Much has been written on this subject in relation to agricultural research in general (e.g. Collinson and Tollens, 1994; Pannell, 1999), and it is not the intention here to enter into a detailed discussion of the issues involved. Assessing impact depends very much on the target group – with the research community, for example, the number of citations of scientific papers is an accepted measure of impact, whereas amongst farmers, the proportion adopting a new technology is often taken as a measure. With educationalists, it may be the numbers of students successfully passing a specific course of study.

The output of modelling is generally information – in practice, however, estimating the impact of information can be difficult (Pannell, 1999). It depends on the perceptions and beliefs of the target group before receiving the information, and the extent to which the information modifies these perceptions and beliefs. We have, therefore, taken impact in relation to crop–soil models as referring to a change in behaviour or thinking of a particular target group as a result of such a model having been used in some way, compared to the behaviour or thinking that would have occurred in the same time frame if the model had not been used. This change in thinking or behaviour should, of course, have some relevance to improving the livelihoods of the poor in developing countries. It is clear that many of the model applications described in this book have had little impact by this definition. Certainly, there has been impact in terms of the numbers of scientific papers published, and many have had some impact in that they provided a useful learning tool for the user or developer of the model. However, many have merely confirmed or quantified what is generally already known, and while others have produced useful information, this often does not seem to have been taken up by anyone beyond the user.

We thought that it might be useful to identify cases in the many model applications that we have reviewed in which, by our judgement, there has been some kind of impact, and look at these in more detail to see if there are particular characteristics that have resulted in the impact. The list is not intended to be exhaustive, and we also recognize that there are probably many cases demonstrating impact of model use in the commercial sector (see Chapter 11) that have not been documented. Nevertheless, we have identified the following examples:

1. *Designing new rice plant types to increase yield potential (Dingkuhn et al., 1991)*: Models helped in defining morphological characteristics of the 'New Plant Type' of rice currently being developed at the International Rice Research Institute (IRRI). (Further details in Section 3.1.)

2. *Designing rice genotypes for weed competitiveness (Dingkuhn et al., 1997)*: A rice crop model was used to identify characteristics important for weed competitiveness in crosses between cultivated rice, *Oryza sativa*, and the wild rice species *Oryza glaberrima*. The results showed that thinner leaves in the vegetative phase and thicker leaves during the reproductive phase was the best combination. Selection trials for this characteristic are currently under way at the West African Rice Development Association (WARDA). (Further details in Section 4.7.)

3. *Evaluating biological N fixation in Namibia (McDonagh and Hillyer, 2000)*: A model describing N flows in crops was used to evaluate if soil N status could be improved through the use of N-fixing legumes inter-cropped with pearl millet in northern Namibia. Results indicated that grain legumes alone are unlikely to be able to improve soil fertility in the region, and that external fertilizer inputs seem to be necessary. The model identi-fied unpromising lines of research, avoiding the waste of scarce research funding. (Further details in Section 4.5.)

4. *Testing of legumes by CARE in Uganda (Keatinge et al., 1999)*: A sim-ple crop phenology model was used to examine the suitability of legume cover species, not grown locally, for hillside regions in Uganda, the results of which were taken up by CARE International in designing field trials. (Further details in Section 5.1.)

5. *Methane emissions from rice agriculture (Matthews et al., 2000c)*: A UNDP-funded project, of which crop modelling was a part, helped to quan-tify methane (CH_4) emission rates from rice fields in five Asian countries, and to evaluate the effects of different mitigation strategies on emissions. The results of the overall project provided field data and knowhow to enable the countries to comply with the stipulations of the United Nations Framework Convention on Climate Change. They should also able to use this data to counter allegations by developed countries that rice agricul-ture is a major source of CH_4, and to offset political pressure from the developed world to reduce industrial development and constrain land use options for resource poor farmers. (Further details in Section 7.2.)

6. *IBSNAT project (Uehara and Tsuji, 1993)*: The IBSNAT project was ini-tiated in 1982 as a worldwide network of crop, soil and socio-economic modellers, and focused on the development and application of a suite of crop models representing the major world crops. A major out-put was the DSSAT software package which contained the crop models themselves, along with a number of programs to facilitate input data man-agement, and analysis and presentation of the model output. The crop–soil models in the DSSAT suite are probably the most widely used models in the world today, despite the project itself having finished in 1994. (Further details in Section 12.4.1.)

7. *Simulation and Systems for Rice Production (SARP) project (ten Berge, 1993)*: The SARP project was a collaborative network of national rice research institutes in South-east Asia, the IRRI, and Wageningen Agricultural University (WAU) in the Netherlands. Its purpose was to devel-op the systems analysis capability of rice scientists in these national insti-tutions to help them in their research. Initial emphasis was on training in the development and use of the models, while later stages of the project concentrated on collaborative research using the models. The network consisted of 16 national agricultural research institutions in nine countries, and by 1990 had trained a total of 91 scientists in systems analysis techniques (ten Berge, 1993). (Further details in Section 12.4.2.)

8. *Shallow irrigation dams in Australia (Clewett* et al., *1991)*: Ten years of experimental data had shown that storing ephemeral runoff in shallow dams for strategic irrigation of grain and forage crops was a viable strategy. However, simulations over 60 years of historical weather data revealed that these results were biased – the experiments just happened to have been conducted in a 10-year period when the rainfall was above average. Use of the model changed the thinking of the researchers before they made recommendations to farmers which would have been misleading in the longer term, and indeed even in the years immediately following the experimental work. (Further details in Section 4.4.)

9. *Early-sown mung bean in Australia (Robertson* et al., *2000)*: A participatory approach involving researchers, farmers, advisers and grain traders was used to explore yield prospects for a spring-sown mung bean crop in October/early November after a winter fallow. Previously, spring sowing was not recommended due to weather damage in the winter leading to low prices in a market demanding high quality, but, with new varieties available, and the opening of a market for intermediate quality grain, these recommendations needed to be rethought. Simulation studies were used to identify possible options, which were then tested with innovative farmers, and the performance of the trial crops compared with benchmarks estimated with the model. Results showed that spring-sown mung bean had a potential for high returns, demonstrated by almost all of the trial farmers saying they would grow mung bean again. (Further details in Section 5.1.)

10. SIRATAC *(Macadam* et al., *1990) and Epidemic Prevention* (EPIPRE) *(Zadoks, 1981) decision support systems (DSSs)*: SIRATAC was a dial-up crop management system developed in Australia to assist cotton growers in making good tactical decisions about the use of insecticides in irrigated cotton on a day-to-day basis. EPIPRE was a similar system developed in Europe between 1977 and 1981 for supervised control of diseases and pests in winter wheat. Both systems were initially very successful and changed the nature of their respective industries. However, farmers rapidly learned from their experiences through subscribing to each system, and eventually did not feel the need to keep doing so. Thus, while the systems became obsolete as operational decision-support tools, they did have impact in helping farmers manage their crops better. (Further details in Section 9.1.)

11. *Deciding on the form of emergency aid to Albania (Bowen and Papajorgji, 1992)*: To assist Albania with an unexpected shortfall in the size of its winter wheat harvest in 1991/92, the United States Agency for International Development (USAID) commissioned the use of the CERES-WHEAT model to decide rapidly whether to offset emergency grain relief supplies with N fertilizer to boost yields with a late-season application. (Further details in Section 9.5.)

Interestingly, with the exception of item 1 below, there appears to be no single factor that is common to all of these examples to explain why there was some impact. In the following, therefore, we discuss various factors that may have contributed:

1. *Competent modellers*: In all of the examples, competent modellers with the ability not only to run the models, but also to modify them for a particular purpose, were involved. In none of the examples, was an 'off the shelf' model (albeit with a 'user-friendly' interface) used by a specialist in another discipline. Our own experience is that rarely does a particular model implementation ever meet exactly the needs of a particular project, and it is usually necessary to 'get into' the model and modify its structure or output for the purpose in mind. Competence in both the model concepts and in programming, therefore, seem a necessary requirement, if impact is to be achieved. This may change, as modelling moves more towards visual environments such as STELLA or MODELMAKER (see Section 12.3), although even then, it is likely that there will still be a need for people who know the physiological mechanisms used in the model, and when and how these can be modified. Even in the case of the SIRATAC and EPIPRE DSSs, the models themselves were run centrally by specialists and not by the farmers who would be using the results.

2. *Working in multidisciplinary teams*: In several of the examples, modellers were working together with researchers from other disciplines, sometimes extension personnel, and sometimes farmers. The modelling was an integral component of the overall project in these cases, not an isolated part 'tacked' on the end as happens with many projects. In the case of designing new plant types for rice at IRRI and the West African Rice Development Association (WARDA), modellers worked together with crop physiologists and plant breeders to take their ideas and test their feasibility and feed the results back to them. In the IBSNAT and SARP projects, model developers worked closely with scientists in various disciplines (e.g. agronomists, crop physiologists, soil scientists, entomologists, weed scientists, water scientists, economists, etc.) who were interested in using the models.

3. *Participatory approach with growers*: A participatory approach with growers would also seem to enhance the chance of impact, as in the mung bean example from Australia. Dissemination of results from research is a general problem (not just related to modelling), but by working with farmers and using models to test their ideas, and discussing the implications of various options with them, the chances of promising technologies being tested and perhaps adopted later should be maximized. Farmers feel that they have some 'ownership' of the ideas, rather than just being told what to do.

4. *Clearly defined problem*: In most cases, the problem was clearly defined, and the answers were not obvious from the beginning. In the rice design example, the problem was defined as a need for genotypes that had a higher yield potential or greater competitiveness against weeds – what was not known was how to go about achieving these. In the legumes in Namibia example, the problem was whether biological N fixation could contribute to maintaining soil fertility in the semiarid conditions there – what was not known was whether grain legumes had the capacity to do this under those conditions. In the case of the CH_4 work, the problem was that rice agriculture was contributing to global warming – what was not known was to what extent, and what could be done about it.

5. *Demand from a target group*: In some cases, there was demand from a target group for solutions to a particular problem. In the example of N fertilizer exports to Albania, policy makers wanted to know what the crop yield response to N applied late in the season would be, so that they could decide to substitute some grain aid shipments for N fertilizer so that the grain could be produced indigenously.

6. *Long-term commitment*: In many cases, there was a long-term commitment to the projects by donors – the IBSNAT project, for example, ran for 12 years, while the SARP project ran for 10 years. Both projects also had considerable budgets. In the case of the IBSNAT project, this had considerable impact as the DSSAT models are probably the most widely used crop models in the world today. Support in the form of a list-server even after the project finished in 1994 has contributed to this continued impact. In the case of the SARP project, modelling expertise by scientists involved in the project has been maintained after its end, although without continued support, this is diminishing gradually (Mutsaers and Wang, 1999). The CH_4 project, of which the modelling was a part, ran for 8 years, while the shallow-dams work in Australia was for 7 years.

7. *Variable environments*: In the case of the shallow-dams work in Australia, the impact of the model came from its ability to put the results from 10 years of field experiments into the context of longer sequences of weather in a variable environment. As it happened, the 10 years of experimentation coincided with above average rainfall conditions, meaning that recommendations from the experiments alone would have been misleading. Even though we have only this one example of the use of models in quantifying risk in such environments, we believe this is one area that models have a large contribution to make – particularly in evaluating the suitability of technologies for semiarid regions commonly found in developing countries.

8. *Quick answers needed*: In the case of the N fertilizer exports to Albania, policy makers wanted to know quickly whether they should substitute some grain aid shipments for N fertilizer so that the grain could be produced indigenously. Within the time frame available, it was clearly not possible to carry out field experiments to obtain the necessary information. The time aspect comes into many of the examples – the same answers could often have been obtained from experimentation, but it would have taken much longer than with using a model. In the N fixation project in Namibia, for example, several years of experiments could have confirmed the results, but the model was able to show that biological N fixation was not a viable alternative unless the quantities of N being fixed were substantially more than is known for tropical legumes. In the testing of cover legumes for Uganda, CARE might have arrived at the same subset of appropriate geno-types for testing after several years of screening experiments, but not within the time frame that the model was able to make recommendations. Similarly, the use of a model to evaluate optimum planting dates for mung beans in Australia gave results that would have taken a number of years to obtain by experimentation alone.

Care must be taken in judging the impact or otherwise of any technology, as other sectors of the R&D sequence are needed to ensure its diffusion. The lack of transfer of information generated by many modelling studies to those that could make use of it is not due to the nature of modelling *per se*, rather it is a general problem relating to dissemination of all agricultural research results. The participatory approach, described in item 3 above and followed in the Australian mung bean example, may go some way to addressing this problem.

Similarly, there is a need to distinguish between short-term and long-term impact. Collinson and Tollens (1994) identify three types of impact of agricultural research in general, which become progressively longer-term in their time scale of effect: (i) a direct impact on production, consumption and human welfare; (ii) indirect impact on the research capabilities of national agricultural research institutions and universities in developing countries; and (iii) a scientific impact which adds to our stock of knowledge on the way things work. The use of models can contribute at all three levels, more so at the scientific level, but also at the institutional capacity level (e.g. the IBSNAT and SARP projects), and at the production level (e.g. the SIRATAC and EPIPRE examples, and the Australian mung bean example). Particularly in the case of (ii) and (iii), attempting to evaluate final impact is of little use in determining the usefulness or otherwise of modelling, as the pathway from diffusion to final impact is too long and there are too many other factors involved. For example, in relation to (ii), developing-country scientists have often said that they benefited personally from training (in general, not just modelling) at the CGIAR centres, but institutional limitations at home prevent them from using their new skills (Collinson and Tollens, 1994). Similarly, in relation to (iii), research on natural resource management usually involves longer-term programmes and less definable products. In many cases, the products will be an understanding of natural processes and the avoidance of losses (e.g. in topsoil saved, intact ground-water, etc.), factors that are not easily quantified for impact assessment, but which are no less important because of that.

14.4 Concluding Remarks

It is clear that the greatest numbers of documented applications of crop simulation models are in the research arena, but unfortunately it also seems that much of this work remains there, perhaps published in academic journals, and rarely reaches the outside world. There are many reasons for this, but essentially they fall into two categories: (i) either the work is not relevant to real-world problems, or (ii) it is relevant, but is not being disseminated effectively. There can be no doubt that a proportion of work in which models were used falls into (i) – often a modelling study merely confirms what was known or practised already. Nevertheless, there is also a significant amount of work involving models which does provide useful information, but which does not reach those who would benefit from it. This shifts, therefore, from being a problem associated specifically with modelling to one associated with agricultural scientific research in general, and which it is beyond the scope of this chapter to discuss.

To help in this process, the criteria for success of modelling activities need to change from a product as an end in itself to a process for achieving outcomes. In particular, there is a need to demonstrate how models can lead to practical innovation rather than just providing analytical justification for knowledge already incorporated into practical models, as is the case with many model applications. There is also a need to recognize both the value of practitioners' models and the limitations of researchers' models, and look for ways and situations in which the latter can improve the former, not just by providing extra precision, but actually providing information that would not be available by quicker and cheaper means. Underlying this must be an appreciation of the practical problems faced by the practitioners – it is no use for example, using a model to recommend to a farmer that he should apply his fertilizer on a specific date if he is not even sure that it will be delivered to his farm in that month or the next, a situation not uncommon in subsistence agriculture in the tropics.

Part 5
The way forward

Where to Now with Crop Modelling?

15

Robin Matthews

Institute of Water and Environment, Cranfield University, Silsoe, Bedfordshire MK45 4DT, UK

In this review we have attempted to document a number of applications of crop–soil models in tropical agriculture, to identify target groups who can potentially benefit from their use, either directly or indirectly, and have discussed model limitations, constraints to their uptake and characteristics contributing to their impact on the agricultural systems they are targeted on. In this chapter, in which we offer some thoughts on the way ahead, we wish to broaden the discussion to include systems analysis methodology in general, not just crop models.

Before we do that, it is useful to put things into perspective, and to look at trends in international agricultural research and where we are at the moment. Rabbinge *et al.* (1994) have identified five stages in the development of international agriculture since the 1960s.

1. Crop improvement through better use of genetic material.
2. Development of technologies and agronomic innovations.
3. Farming systems research, tailored to farmers' needs.
4. Awareness of environmental side-effects and the need to contain them.
5. Integrating strategic, basic, applied and participatory research and tailoring it to ecoregions.

We are currently at Stage 5, in which knowledge generated in the previous stages is being integrated and used within ecoregions. Related to this is a shift in thinking away from a focus on natural resources and commodities only, towards a more 'people-centred' approach which considers the totality of the ways in which people make their livelihoods. This thinking is represented by the Sustainable Livelihoods (SL) framework (Fig. 15.1), which is discussed in more detail in the next section.

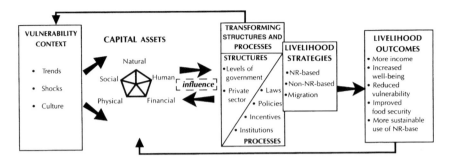

Fig. 15.1. The Sustainable Livelihoods framework (from Carney, 1998). NR, natural resources.

Models should be seen as vehicles for the encapsulation and transfer of the knowledge gained from the many field and laboratory experiments carried out in the past, and a way in which this knowledge can be used to understand and predict the behaviour of natural systems, without the need to repeat the same research all over again at different sites. Seen in this way, models fit neatly as essential tools into the current phase of international agricultural research described above. In doing so, they are capitalizing on the past investment in research and adding value to current research expenditure.

In the following sections, we suggest ways in which modelling approaches can contribute to developing agriculture. We see six broad areas emerging:

- research on how people influence, and are influenced by, the biophysical world, including changes in climate;
- research on linking genome and phenotype, and how this may contribute to crop improvement programmes;
- using models to help develop decision support systems to aid in the dissemination of existing knowledge;
- using models of crop and soil processes to understand and predict the interactions between global change and food production systems;
- integrating use of existing models into projects, where possible, to make the best use of past research and add value to current research;
- maintaining and expanding existing modelling expertise in developing country institutions by supporting existing model users.

These are discussed in more detail below.

15.1 Modelling Rural Livelihoods

15.1.1 The SL framework

Several international development organizations are currently promoting the use of the SL framework as a way of thinking about the objectives, scope and priorities for development in order to enhance progress on the elimination of poverty (Ashley and Carney, 1999). This thinking has arisen from International Development Targets to halve the proportion of people living in poverty by 2015, and has grown from the recognition that it is fruitless to try to solve technical problems without, at the same time, addressing the socio-economic pressures against which they are set. Human behaviour, driven by poverty, population dynamics and inappropriate economic and land policies, is usually the underlying cause of land and water degradation. In turn, human behaviour is not always driven by a desire to maximize yields or income – these may be only two objectives out of many – other goals may include income stability, product diversity, an attractive landscape or concern for family.

The main feature of the SL approach is that it places people at 'centre stage', rather than natural resources or commodities as has been the case in the past, and considers their assets (natural, human, financial, physical and social capital) and their external environment (trends, shocks and transforming structures and processes, see Fig. 15.1). A key concept is that of 'sustainability' – a livelihood is defined as sustainable where it can cope with, and recover from, stresses and shocks, and maintain or enhance its capabilities and assets both now and in the future, while not undermining the natural resource base (Carney, 1998). As such, the SL framework challenges development professionals to think about the whole livelihood system rather than just some part of it, and may, therefore, herald a significant evolution in conceptualizing key processes and relationships in rural development.

However, at present it is not clear as to the extent to which the SL framework may be used in a diagnostic sense (i.e. to identify the real constraints of a particular system), to determine a course of action, or to test out the likely outcomes of particular strategies. Further thought is required to make explicit the various components of the current SL framework and to enable them to be explored in a rigorous and testable way.

15.1.2 Can modelling contribute to the SL approach?

The one characteristic that clearly distinguishes agricultural systems from any natural system is the human element – all agricultural systems are based on humans who manage natural systems to meet their objectives. Despite it being self-evident that it is the human element of farming systems which determines the success of policy and technology transfer initiatives, relatively little research has gone into understanding and predicting the dynamics of the social elements of agricultural systems (Edwards-Jones *et al.*, 1998). Some progress in this direction has been made in the last 25 years, but social systems have not been analysed as intensively or as quantitatively as livestock or crop production.

The importance of developing models of farm households and linking these with biological crop models in a whole-farm model was recognized by Dent and Thornton (1988). In a subsequent paper, Dent (1993) suggests that such a model could be used to evaluate the likely impact of new circumstances that might include alternate new technologies, alternate market scenarios and credit facilities. Adoption rates of a specific technology could be predicted, as well as total farm output and its stability. Alternate technologies could be screened before experimentation, and judgements could be made about the relative merits of technology, education, credit facilities and information provision. Such a model would need to be generic in nature so that it could be applied to different types of households represented by different parameters, but without structural modification of the model itself. It would also be necessary to devise some sort of minimum data set of socio-cultural and economic data.

Since then, however, relatively few models which attempt to simulate the behaviour of farm families or to integrate their behaviour with other elements of the agricultural system, have been developed (Edwards-Jones *et al.*, 1998). Jones *et al.* (1997) suggest three levels of farm-system models.

- *Unconstrained*: Models that focus on the ecological components only. An example of this type of model is that developed by Hansen and Jones (1996).
- *Resource-constrained*: Models that incorporate the economic component in addition to the Unconstrained level, i.e. what is economically attainable if one or more resources are fixed. This level of model may include a simple decision-making component based on maximizing profits. An example of this type of model is given by Edwards-Jones *et al.* (1998) who linked CERES-MAIZE and BEANGRO to a family decision-making and demographic model to represent a subsistence farming system. However, the work highlighted a number of issues that need to be addressed, which were discussed in more detail in Chapter 5.
- *Adaptive*: Models which incorporate the social component in addition to the Resource-constrained level. These should include household characteristics and preferences that influence decisions such as acceptance of new technology and choice of crop and livestock enterprises. Account should be taken of farm goals, which are probably more complex than just profit maximization. An example of this type of model is given by Dillon *et al.* (1989) – this included a decision model that considered

farmers' risk preferences. The FLORES (Forest Land Oriented Resource Envisioning System) model developed at a workshop in Indonesia in February 1999 is another step in this direction. Generally, however, little progress has been made in this area due to lack of a mechanistic understanding of household decision-making processes.

The basic assumption underlying the theoretical framework of farm household modelling at the Adaptive level is that decisions on land use are taken by individual households based on their goals and aspirations, making use of the various available resources to undertake specific activities subject to biophysical and socio-economic constraints (Kuyvenhoven *et al.*, 1995). A complicating factor is the existence of many different households with different objectives, which makes it necessary to distinguish between groups of households that show broadly similar reaction patterns. The interest is in the overall response of farm households to policy change, not the reactions of a specific household. Kuyvenhoven *et al.* (1995) suggest classification on: (i) resource endowments (e.g. land/labour ratio) and (ii) differences in objective functions.

Thus, we would suggest that one approach that can be taken towards the further development of the SL framework is through the modelling of livelihood systems, which, when combined with geographic information systems (GIS) and remotely sensed data interpretation techniques, should have the potential to provide a powerful tool to evaluate the dynamics within particular communities, the likely effect of outside influences and the range of livelihood strategies that may be adopted to cope with these. Existing crop–soil models would be an integral part of such livelihood models, particularly in the case of rural communities. By combining the strengths of both biophysical and socio-economic modelling, a more effective systems research approach for the development of improved farming strategies should emerge. Biophysical modellers must realize that their areas of speciality are just subsystems within much larger systems in which people make their livelihoods. Improving such systems involves the recognition that several, perhaps conflicting, goals need to be satisfied. It may even be that their own subsystem is not very important in the overall picture. Social scientists, on the other hand, must realize that models offer an objective and testable framework not only for understanding a system, but also for examining the consequences of manipulating it or intervening in it. Both disciplines have much to learn from each other and of each other's tools.

The concept of the 'five capitals' (human, social, financial, physical, natural) is central to the SL framework, and any modelling effort must address these. In the following, we suggest some initial ideas of how these five capitals could be simulated within an SL model:

1. *Human*: This includes skills, knowledge, ability to work, etc. It is envisaged that this could be described initially by a simple variable for each individual representing the degree of skill that they have – this could be related, for example, to the number of years spent at school. At some stage, this may have to be subdivided further, say, into experience built up over years and the ability to carry out work of a particular type – an old person, for example, may not be able to do much physical work, but may have considerable experience that could be passed on to younger members of the household.

2. *Social*: We see this as a function of relationships between individuals, but mediated through social institutions. Each individual (within and between households) would possess information about the number and nature of linkages between themselves and other individuals. Key social relationships to be explored would include kinship (ties both within and between households), gender and friendship. The 'value' of key relationships would need to be assessed, possibly with reference to different notions of reciprocity which may be useful in differentiating kinship from other aspects of an individual's social capital.

3. *Financial*: For each household, a variable representing a money store could be defined, which can be augmented by income generated by members of the family and diminished through expenditure by members of the household. A similar variable could represent the size of the food store, which would be added to by crop harvests or purchase of food and subtracted from through consumption by members of the household.

4. *Physical*: This includes infrastructure such as roads, schools, markets, etc., and would be represented spatially within the GIS. Proximity of households to these is likely to influence the household's decision-making processes.

5. *Natural*: This includes climate, soils, land and water information, databases of which exist for many areas already and are routinely used by current crop–soil models. Where such data do not already exist, much could be derived from the interpretation of the remotely sensed data in a participatory framework. All natural capital data could then be stored in a GIS format for use with the household model. Crop–soil models provide one of the links between the household model and these aspects of the natural capital.

It would seem appropriate to start modelling at the level of the household, as several of the capitals (i.e. human, financial and natural) operate at this level. Once this was achieved, a number of households could be aggregated to make a 'virtual community', and the social and economic interactions between them incorporated. This virtual community would then be embedded in an 'environment' consisting of climate, fields and physical infrastructure. Each household would, in effect, become an 'intelligent virtual agent' capable of sensing its environment and interacting with other households, and being able to change its actions as a result of events in the process of interaction. Emergent properties, such as price levels for different commodities, would be a result of the aggregated behaviour of individual households, in this case, the supply and demand of the commodity and its elasticity. The whole system would be similar to a multi-agent system (MAS; Sichman *et al.*, 1998), which we discuss in Section 15.4.

Such an SL model would be a simulation model as opposed to an opti-mization model. Most household models to date have traditionally used some kind of linear programming (LP) approach, which finds an optimum mix of resources in order to meet specified objectives (e.g. maximizing income, minimizing risk). The limitation to this approach is that it pre-supposes a 'goal' of the household, and does not adequately consider the day-to-day decisions being made by householders. However, most human action is driven by routine reaction to the peculiarities of a situation rather than by elaborately calculated plans or goals (Brooks, 1990). Also, such models are structured to represent an equilibrium, and represent a time when production has stabilized. Climate conditions, for example, are assumed to be average every year. They, therefore, are not able to simu-late processes leading from one equilibrium to another particularly well. The approach we are proposing is dynamic, and as such, it would be a tool for exploring the implications of different options rather than recom-mending a specific option as being the 'best'.

Once the SL model, incorporating explicit representations of the differ-ent capitals, had been developed, it could then be used to investigate the vulnerability of different communities and the different types of households within them to outside influences, such as climate variability, policy changes, etc. Household (or community) vulnerability can, to a large extent, be related to the total stock of capital available to it in times of crisis. However, it is not necessarily the case that a household with a greater stock of capital will be less vulnerable than those with less capital. An important characteristic influencing the ability of a household to with-stand outside shocks is the ability to convert capital from one type to another. For example, social capital built up over the years in the form of linkages between individuals could be converted into financial capital in times of adversity (i.e. a cash transfer between relations or friends). Not all forms of capital can be converted into others – for example, physical cap-ital (e.g. roads) cannot easily be converted into natural capital. Some forms of capital may be easily converted to another, but not easily reconverted back again (e.g. natural capital could be converted into financial through cutting down trees, but could not quickly be converted back into trees again). The SL model could be used to explore the importance of all of these conversions.

The SL model would, of course, be a research tool in the first instance, to help explore and test the relationships between the concepts currently incorporated in the SL framework. As such, it would be an important part of any further development of the SL approach. Nevertheless, after some time, it is envisaged that specific tools would be developed from the model, which practitioners would be able to use to help them in designing their projects with a greater SL focus. By understanding the socio-economic and biophysical processes of the system better, it should be easier to design pathways out of poverty, i.e. given that it is rational behaviour for a household to want to enhance its livelihood through food security, cash generation and quality of life, can ways be found or policies devised whereby this be achieved without degrading the environment? Possible functions these tools might provide include:

1. *Evaluation of new techniques*: An important contribution an SL model could make is in screening out options that can be demonstrated to have a poor chance of success, rather than wasting scarce research funds on them. An example of this was the evaluation of grain legumes for improving soil fertility in Namibia (McDonagh and Hillyer, 2000) described in Chapter 4. Alternatively, they can be used to identify candidate techniques not currently being practised in a region, which may have been overlooked in the traditional farming systems research (FSR) process, such as was the case in one project in Kenya (e.g. McCown *et al.*, 1994). The SL model would allow them to be seen in the context of the whole farming system alongside existing enterprises, rather than as isolated techniques.

2. *Extrapolation from specific sites*: An SL model could also be used to extrapolate results from a limited number of study sites to wider areas, and help to identify 'extrapolation domains'. The concept of using locally validated detailed models of farming subsystems as the preferred medium for applied research has been discussed by Dent and Edwards-Jones (1991).

3. *Provision of a temporal dimension*: The current SL framework provides a 'snapshot' of a current system, but does not easily allow an analysis of future changes to the system. An SL model can be used to provide an evaluation of the short-term risk faced by households and also of the long-term sustainability of various interventions. Cumulative probability functions can be used to assess the variability in household income or harvests between years. Complex strategies, such as cropping sequences with a range of management rules, could be explored in this way, to evaluate stability of output and economic achievement.

4. *Providing mechanisms to predict the effect of outside influences on the vulnerability of livelihoods*: An SL model could provide a way of objectively assessing the impact of outside influences (e.g. government policies, commodity prices, environmental impacts) on various components of a system, including the livelihoods of the people within it. For example, the question of how the building of a road through an area affects land use, cost and availability of food, household vulnerability, etc. could be explored by such a model.

5. *Providing mechanisms to predict the effect of improvement in livelihoods on the environment*: In the same way that the effect of external influences can be explored, an SL model could enable an assessment of the environmental impact that an improvement in livelihoods may have. The environment in this case includes both that within the system (e.g. the effect of intensification on runoff into rivers) or external to it (e.g. the effect of rice cultivation on global atmospheric methane levels). Spedding (1990) has pointed out that those who advocate change have an obligation to consider the full consequences of making the change, not just in terms of the objectives sought, but also any possible undesirable consequences.

Some initial case studies should be carried out to test and validate the model, and also to demonstrate the potential usefulness of the approach. Some suggestions could include:

- The effect of population increase on the length of the fallow period and on long-term soil fertility in shifting cultivation systems. Are improved fallow systems justified in such systems?
- The effect of agricultural intensification on livelihoods and on the rates of deforestation.
- The effect of migration to urban areas on the sustainability and productivity of traditional cropping systems.
- The effect of external forces (e.g. climate change, government policies) on the vulnerability of livelihoods of households of differing resource levels.
- The effect of building a road in a remote area on livelihoods and land use within that area.

These are only a few examples, and many more can be thought of.

15.2 Contribution to Crop Improvement Programmes

The use of crop models in crop improvement programmes is in its infancy, but their potential in helping to identify and evaluate desirable plant characteristics, environmental characterization, and explaining genotype × environment (G×E) interactions (see Chapter 3) needs to be explored. Recent advances in biotechnology are also opening up new opportunities for crop improvement through such techniques as molecular marker assisted selection and genetic engineering for a wide range of crop species, and there is, therefore, likely to be a crucial role for crop simulation models to play in linking information at the gene level to that at the phenotype. However, it is not clear as yet how this should be done – the big gaps in our knowledge are the mechanisms whereby proteins are formed from their constituent amino acids. In the long term, the emerging field of proteomics may provide answers in this direction, although the indications so far are that enormous computing power is required to simulate the processes of folding during protein formation. On the other hand, classical quantitative genetics has always relied on statistical relationships between the effects that the alleles of genes have on the phenotype of individuals, and ultimately populations of individuals, without the need to consider the biochemical pathways involved. It is likely, therefore, that considerable progress in our understanding can be made by linking allele effects directly to phenotypic responses and using this as input for models of plant breeding systems, such as was attempted by Chapman *et al.* (2002). More precise definition of the actual genetic architecture of various phenotypes seems to be a logical next step, building on the initial simplistic assumption made by these authors of two-gene/two-allele systems for each of the traits. Other attempts to use crop simulation models to link phenotypic characteristics to quantitative trait loci (QTL) have demonstrated some potential (e.g. Yin *et al.*, 1999a, b), but have also underlined some of the inadequacies of the models in this respect (Yin *et al.*, 2000).

Once suitable models linking gene to phenotype have been developed, ideally they should be fully integrated into plant breeding software systems. Efforts are already underway to achieve such integration for the International Crop Information System (ICIS) being developed by some of the CGIAR centres (Fig. 15.2). White (1998) has suggested that crop models able to directly simulate gene action for specific processes will be able to provide a link to genome databases, enabling molecular biology techniques

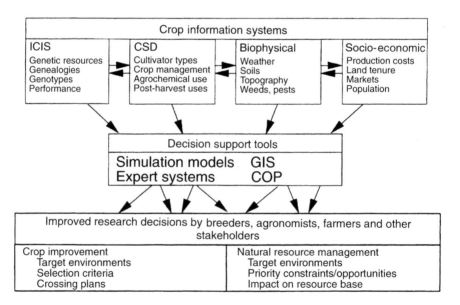

Fig. 15.2. Possible structure for a software system for plant breeding, including crop simulation models and geographic information systems (GIS) (from White, 1998). ICIS International Crop Information System; CSD, Cropping Systems Database; COP, Coefficient of Parentage (routines in ICIS).

to characterize genotypes so that the reliance on field trials to fit model coefficients can be reduced. Indeed, in most situations, experimental evaluation of molecular breeding strategies for manipulating complex yield adaptation traits will be impractical or beyond the resources of breeding programmes (Chapman *et al.*, 2002), and it is likely that modelling tools will play an important role in this area. How feasible all of this is, and over what time period it can be achieved, remains to be seen.

In many cases, models may not be able to make much contribution in determining which characteristics a genotype should possess. Where a particular trait such as resistance to a specific disease is clearly required, there is little more to be done than for breeders to go and incorporate that trait into the appropriate genotype. Models can make their greatest contribution if there is a cost or trade-off associated with a particular beneficial trait – for example, if the trait for disease resistance meant that in years when there was no disease, yields from a genotype containing it were lower than for a genotype without the trait. A model can then be used to identify the environments in which it is advantageous for genotypes to possess the trait, and also assess the risk to a farmer over a period of time of growing such a genotype.

15.3 Making Information Available – Decision Support Systems

In the SL framework discussed above, knowledge is seen as part of the 'human' capital. In relation to improving livelihoods of poor people, the knowledge of how to do so may often exist (i.e. from formal research or indigenous experience). However, often the bottleneck to it being successfully applied to solve problems is making it available in an understandable form to those that can make use of it, a theme that we have reiterated a number of times in this book. Provision of decision support systems (DSSs) is one way in which this can be done.

Having reviewed a range of DSSs, their application and some of the reasons why this has been limited, it is useful to consider the ways that their potential can be optimized. Chapter 11 highlighted some general issues which can be used to determine practical guidelines. The main problem is that models have not met the needs of end-users – the way forward, therefore, is to take more deliberate steps to meet these. If decision support models are to be useful, then they must be based on pragmatic rather than scientific criteria. They must address real problems faced by the end-users rather than problems of interest to the researchers. This requires early definition of the end-users. Models for DSSs should only be undertaken when they are appropriate to the problem and can realistically produce outputs that are usable by the clients. It is important that before any DSS is developed, the problem has been rigorously defined and the requirements of the decision maker are fully understood (Knight, 1997). End-users should be involved in the development process. There is a need for effective linkages between researchers, extension workers and farmers (Hamilton *et al.*, 1991) as well as between researchers and policy makers. As already discussed in Section 15.1, one way to increase the applicability of models is to combine 'hard' and 'soft' systems approaches (Newman *et al.*, 2000) – in other words, a more participatory approach to DSS design and development, involving representatives from research, extension and farmers' groups, is required (Box 15.1).

Box 15.1. Requirements for successful DSS packages.

1. Address real problems (often complex) not readily solved by rule of thumb (e.g. pest management and irrigation scheduling require decision making on issues that vary from one season to the next. The cost of making a mistake is high and therefore use of DSS may be worthwhile. Time of fertilizer application is less important as the responses are flatter).
2. Address problems that will be costly if the decision is not made correctly (i.e. where there is a steep response).
3. Must be easy to use and output easily understood.
4. Must be targeted at the client.
5. Must not require an experienced computer programmer to operate, or must be part of a system where the operator works as a consultant passing on the relevant outputs in a useable manner.
6. Must be introduced to the client with a thorough training package and continued support.
7. Need to be maintained and updated with changing technology and in response to user demand.

All systems should be developed as part of a project that includes a budget for dissemination and training. Newman *et al.* (2000) suggest that the DSS must address significant problems identified by the users in the short term and simultaneously adopt a longer-term strategy towards development of educational software for training in the next generation. Evaluation, maintenance and upgrading of the system should be incorporated into the model development project. The marketing and dissemination strategies should be considered at the start of the process. Training should not be compressed into too short a time period. A combination of formal training and extended on-the-job programmes was found to be the best way to ensure that procedures were fully understood and adopted. It is also important to train staff to collect the appropriate data in the field (Makin and Cornish, 1996). Training should focus not only on how to use the system but also on the assumptions that are involved in model development and the potential and limitations for their application. It should be highlighted that the models are aids to decision support and not ends in themselves. Care should be taken not to promote a computer-based decision aid approach when other communication means are more appropriate (e.g. the written word). Accessing related information, analysis techniques, local experience, financial implications, etc. should all be seen as part of the package.

From a development perspective, a DSS would be most appropriate where considerable research on identified constraints to improved livelihoods has being going for some time. For example, such an area might be the mid-hills of Nepal where much research on soil fertility and soil conservation has been carried out over the years. It may be time to take stock of all the information that has been gathered from this research, and attempt a synthesis of it into a form that could be used by practitioners that have to make decisions on the most appropriate practice to recommend. Both biophysical and socio-economic information should be incorporated. Crop–soil models should be used wherever possible to derive simple rules of thumb that can be incorporated into the DSS. An example of this might be the relationship between altitude and the rates of organic matter build-up or N fixation through the influence of temperature (see Section 15.7.2). Existing farmer knowledge should also be taken into account in the development of the DSS. New areas of research may be identified as a result of this exercise. The most appropriate form of the DSS should be determined together with those who will use it, but it could be one of, or a combination of, a computer-based system, flipcharts, handbooks, manuals, etc.

15.4 Integrating Model Use into Research and Extension Projects

We also believe that models have a valuable contribution to make to larger projects in being able to explore various scenarios and management options, particularly from a long-term perspective. It is important that this is done in a participatory way with the stakeholders, whether they be researchers, extension personnel or farmers.

The degree of participation will vary with the end-user – other researchers, for example, may be keen to build their own models, or add to an existing one. In such cases, commercial visual modelling packages, such as STELLA, MODELMAKER, in which models are constructed visually on a 'box-and-flow' basis without the need for programming ability in a computer language, may be the most appropriate approach. There is a tendency with these packages, however (as with all models), for the models built in them to become increasingly complex to the extent that they too become incomprehensible to other users, and also model performance may be severely downgraded (as was found with WALNUCAS in STELLA; G. Cadisch, Wye College, Ashford, 2000, personal communication). Future developments in visual modelling packages should address these problems.

Other researchers, and perhaps consultants and agricultural extension personnel, may not wish to build or modify models at all, but prefer instead ready-made models, linked to 'user-friendly' graphical interfaces which facilitate their use. Subsistence farmers, on the other hand, are unlikely to be direct users of either of these types of models in the near future, not least because of the general unavailability of computers they would have access to. Nevertheless, they may be very interested in the predictions of such models, so that the most appropriate approach may be for competent modellers to act as the interface between them and the model.

It is interesting to note that in nearly all of the examples of model impacts discussed in Chapter 14, competent modellers were involved. Modellers should, therefore, be an integral part of the team involved in the overall process of livelihoods improvement, and it is important that they also have the opportunity to enter into dialogue with farmers and other target groups. Such an approach is in the process of being developed by the Agricultural Production Systems Simulator (APSIM) group in Australia (McCown et al., 1994). Clients (not just farmers, but all decision makers in agricultural development) are involved in projects. If the clients are farmers, collaborative experiments are done and results are extrapolated in time using models to show the long-term consequences of their actions. If the clients are researchers, models are used for extrapolation of the research results in space and time to add value to expensive research. The approach described by Robertson et al. (2000) in developing new cropping strategies for mung bean in response to changed external factors in Australia is a good example of how this can be achieved (see Chapter 5 for details).

Another example of how the use of crop–soil simulation models within a team can help to provide insights into ways current farming systems (and hence rural livelihoods) can be improved, is provided by the same group (McCown et al., 1994). They describe the development in a part of Kenya of a downwardly spiralling 'poverty-trap' due to increasing population pressure, leading sequentially to nutrient depletion, yield reduction, income reduction, soil degradation, and eventually to low crop yields and low income (Fig. 15.3a). A small number of on-farm field experiments backed up by modelling showed that use of small amounts of fertilizer (40 kg N ha^{-1}) was an efficient strategy for breaking the poverty cycle and long-term improvement of the system (Keating et al., 1991), even taking into account risk (Fig. 15.3b). Extensive FSR had previously not considered this as an option as 'farmers here don't use fertilizer'.

A similar approach is currently being undertaken in Zimbabwe by the same group (P. Grace, CGIAR, 2000, personal communication) in which extension staff and researchers were invited to a workshop together, and models were used to explore trade-offs between production and risk for

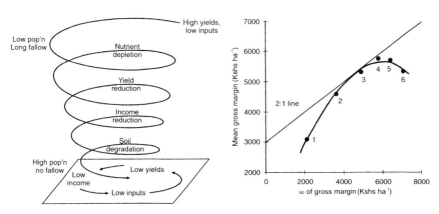

Fig. 15.3. (a) The poverty trap that results from human population pressure. (b) The values of various maize production strategies in Kenya depicted in mean–standard deviation space. Strategies combined planting densities (plants ha^{-1}) and fertilizer N (kg ha^{-1}). Points represent (1) 22,000 plants ha^{-1}, 0 kg ha^{-1}; (2) 27,000 plants ha^{-1}, 15 kg ha^{-1}; (3) 33,000 plants ha^{-1}, 30 kg ha^{-1}; (4) 38,000 plants ha^{-1}, 45 kg ha^{-1}; (5) 44,000 plants ha^{-1}, 60 kg ha^{-1}; (6) 55,000 plants ha^{-1}, 80 kg ha^{-1} (from McCown *et al.*, 1994).

various scenarios suggested by the extension staff. The workshop was considered successful in that it highlighted to the researchers the challenges that small-holders face when investing in fertilizer inputs and provided some pointers for future on-farm research.

An interesting development is in the use of MAS modelling and role-playing games in the management of natural resources. A modelling paradigm originating from the field of distributed artificial intelligence, MASs are potentially suitable for linking the biophysical and socioeconomic characteristics of a system. These models consist of a number of 'intelligent' virtual agents (IVAs) which are sensitive to other agents and interact with both them and their environment, and can change their actions as a result of events in the process of interaction. Barreteau *et al.* (2001) describe the application of a MAS model to irrigated systems in the Senegal River Valley. These irrigated systems were built less than 30 years ago, but have given disappointing results since then. The model was used to investigate a possible link between this poor performance and stakeholder coordination. Three groups of farmers were defined – one for allocation of credit, one for allocation of water and one for pumping management. Each group was autonomous with its own set of rules on how to behave, but could communicate and interact with the other groups. In addition to belonging to one or more of these groups, farmers could also act as individuals. At the beginning of each season, a farmer would need to decide what to grow, if anything, how much water he required and how much credit to apply for. His request would be considered by the water allocation group, and if successful, water would be pumped to him by the pumping group, provided certain criteria were met. The MAS model was converted into a role-playing game with the purpose of letting the participants (i.e. the actual farmers involved) see the complexity of managing a renewable resource, and the absence of a single answer to the question of how this resource is to be managed. They found that the game and the model were powerful tools for initiating discussion amongst the stakeholders regarding their collective behaviour in the system. Several of the players requested that they keep the game with them to use in their own villages to support discussions about their irrigated systems.

In summary, the use of models is an efficient and cost-effective way of exploring possible options under different scenarios, and could be used as a way of testing proposed interventions where appropriate, particularly in areas where it is difficult to carry out field experiments due to the nature of the terrain or to remoteness. It might be a useful exercise for those planning future projects to ask themselves if any of the work they are proposing could be answered by existing crop–soil models rather than carrying out expensive, and possibly unrepresentative, field experiments. Perhaps there should be a box in the project application form indicating whether the applicant had consulted a modeller in the same way that it is often required to do so for a statistician! We do not advocate the complete substitution of field experimentation by modelling, as there is always the need for validation of the models, but the amount of experimentation, both in terms of the numbers of replications at different locations, and in the numbers of years, could be substantially reduced. Modelling can be used in this way to add value to the experimentation as well as capitalizing on the investment already made by funding agencies.

15.5 Environmental Research

Environmental research in developing countries is clearly one area in which crop–soil models can make a major contribution. The advances made in crop production over the last three decades have not been without cost to the environment – N applied through fertilizers, for example, has increased more than seven times, and is beginning to pollute many rivers and water tables. Moreover, there is evidence of diminishing gains from these inputs due to land degradation caused by many factors, including poor land-use planning resulting in deforestation and clearance of marginal land for cultivation, poor management of water resources and agricultural land, over-use of pesticides and fertilizers, uncontrolled dumping of wastes and deposition of pollutants from the air (UNEP, 1999).

Because of the predicted trends in global population, there is a definite need for further gains in agricultural productivity, but this must be achieved in a sustainable and environmentally friendly way so that the natural resource base is preserved. Also, because of the multifaceted nature of environmental problems, the emerging challenge is to develop ways of linking science, social and environmental data in meaningful ways to help achieve these goals. Two areas in particular in relation to agriculture were highlighted in a recent report by the United Nations Environmental Program (UNEP, 1999):

- *Climate change*: The balance of evidence now suggests that there is a discernible human influence on global climate (IPCC, 1996). While many of man's activities are involved, agriculture is an important one, both because of the influence it has on the emission of greenhouse gases into the atmosphere (e.g. CH_4 from rice cultivation), and also because of the impact that a changed climate is likely to have on agricultural production and its ability the meet the demands of the expanding population mentioned above. In a recent poll of 200 environmental experts (UNEP/SCOPE, 1999, cited in UNEP, 1999), climate change was the issue for the 21st century mentioned the most by the respondents. Future work in this area should focus on integrating the results of large number of studies of different components of climate change. For example, in relation to rice agriculture, three possible areas of research are:
  - Rice production is influenced by increasing CO_2 and temperature, but in turn influences CH_4 emissions into the atmosphere. Nitrous oxide, another greenhouse gas, is also emitted from flooded rice fields under certain conditions. Studies are required to investigate which crop management options result in the least detrimental effect overall to the environment taking into account all of these factors.
  - Another important study is on how sea level changes under a changed climate are likely to affect overall rice production through their effect on rice growing areas.
  - A third important area of research is developing ideas on how people relate to all of this – how decisions made at a household level collectively contribute to changes in land use over time, and how this influences, and is influenced by, climate change.

  Similar areas of work could be identified for other cropping systems.

- *N in the environment*: Evidence is mounting that human activities are seriously unbalancing the global N cycle. The advent of intensive agriculture, fossil fuel combustion and widespread cultivation of leguminous crops has led to additional quantities of N being deposited into terrestrial and aquatic ecosystems. Surprisingly, given the importance of the process, there are few models for biological N fixation, and one area of research should be the development of a process-based model describing potential N fixation rates in relation to environmental factors such as temperature, water status, host plant growth, etc. This point is discussed in more detail in Section 15.7.2 below.

Future research needs to consider carefully the impact any recommended interventions might have on the environment in the light of these two effects. Crop–soil models are important tools in helping to make these assessments, and the SL modelling approach described above is one way of integrating these with the human dimension.

15.6 Building Modelling Capacity

The argument that developing modelling capacity is beneficial in less-developed countries (LDCs) is seductive. This would break the chain of modelling dependency on developed countries and help to develop an internal capacity for modelling that can better answer the needs of LDCs. Important skills could be transferred rather than finished products. However, the question as to how useful it would be to develop modelling capacity in LDCs needs to examine whether mechanistic or functional modelling capability is to be developed.

There is no reason why, with the use of visual modelling tools or spreadsheet packages, the development of modelling capability in simple functional models should not allow researchers, decision makers or tutors in developing countries to further their work. To a certain extent this is already happening – simple models to examine the effects of water abstraction on groundwater levels in southern India, for example, have been developed with Excel and Visual Basic (Ravi Bhalla, Pondicherry, India, 2000, personal communication). This capacity for the production of simple functional models could be usefully developed. Additionally, this kind of capability might be developed with the computing skills and finances that are already available, an important consideration where both are limited.

However, the enhancement of the mechanistic modelling capacity in LDCs is a process that is beset with problems. Such modelling capacity is difficult to transfer because mechanistic models are extremely complex. The successful development of mechanistic models may require international and interdisciplinary teams working with professional modellers rather than agricultural scientists who are taught to model. Assembling such teams is expensive and in view of this it is probably more cost effective to modify existing models where appropriate rather than to develop modelling capability. Mutsaers and Wang (1999) have shown that there was only limited independence in modelling skills gained by National Agricultural Research Centres (NARCs) in LDCs after the Simulation and Systems Analysis for Rice Production (SARP) project finished. The modelling skills developed during the project are in danger of being lost without further support (see Section 12.4.2).

Development of user support groups to provide help and model updates to users of existing models is one way of trying to encourage the maintenance and development of these skills after the main project has finished. The emphasis in such support groups should be on the applications of the appropriate models to solving practical problems of importance in the research areas of the members. For example, agroforestry models could be used to identify conditions and regions in which alley cropping is likely to be successful, taking into account both the biophysical and socio-economic aspects that are influential.

15.7 Further Crop–Soil Model Development

There is no doubt that crop–soil simulation models, describing crop growth and yield in relation to climate, soil and management, have reached a stage where they can provide useful insight into complex cropping systems, although, as was shown by the discussion of various model limitations in Part 1, they are by no means perfect yet.

It would seem sensible, however, in the current context within international agriculture of applying knowledge gained from previous research to solve specific problems, that the emphasis should be more on using existing crop–soil models to obtain answers to appropriate questions and less on further development of such models. This is not to say that further development of crop–soil models should cease altogether, but more that the funding of projects in which the development of a model is the main output are increasingly difficult to justify. Further enhancements to existing crop–soil models could be made as a subactivity to a main project with clearly defined problem-solving objectives, and indeed, this is usually necessary, as often existing models are not able to address a specific problem entirely. This was found, for example, in trying to simulate the dynamics of CH_4 emissions from rice fields, in which it became necessary to link a crop simulation model to a soil CH_4/O_2 model (Matthews *et al.*, 2000a) as neither model by itself was adequate to evaluate crop management strategies that could be employed to reduce CH_4 emissions.

For completeness, therefore, in the following sections, we discuss some areas in which we feel crop–soil models still need further development.

15.7.1 Linkage to pests and diseases models

Pests and diseases are responsible for large crop losses in the developing world, and it is important to understand the processes involved in their population dynamics and how they affect crop yields. Some progress has been made in linking pest models to crop models (see Chapter 4) but most of these combined models still require information on the pest dynamics as an input. Further work is needed to link crop models to models that can predict pest dynamics in a mechanistic way. Such models will have the potential to explore yield consequences of various pest infestation scenarios across a range of pest situations and cropping conditions, including single and multiple pests, varying patterns of pest progress, and varying crop ages, all under different environments and cropping practices

(Teng et al., 1998). Economic factors and analysis of risk also need to be taken into account. The results should allow the prediction of thresholds indicating when particular control measures should be taken, the development of 'least-loss' strategies, and the optimization of pesticide application schemes. Such information from extensive simulation studies can be presented in a condensed form such as simple regression equations, iso-loss curves, least-loss look-up tables, etc. (Teng et al., 1998). These tools could then be used in the field by practical decision makers.

15.7.2 Soil processes

An understanding of soil processes is increasingly recognized as funda-mental to the maintenance of sustainable livelihoods, particularly in eval-uating the long-term implications of various livelihood strategies.

For example, the dynamics of soil C and N processes are fundamental to soil fertility and crop productivity. However, the original soil N trans-formation module used in the Decision Support System for Agrotechnology Transfer (DSSAT) and other crop models is based on a submodel (PAPRAN, Seligman and van Keulen, 1981) originally designed for high-input agricultural systems, where soil organic matter (SOM) is generally not considered of great importance for the nutrient supply of a crop. Some of these limitations have now been addressed by the substitution of routines from the CENTURY SOM model by Gijsman et al. (1999), although there may still be limitations in highly weathered soils (Gijsman et al., 1996). Work is also currently underway at IACR Rothamsted to develop SOM routines specifically for flooded rice soils using C pools that can actually be measured rather than arbitrary ones that most models use (Gaunt and Arah, 2001). It is hoped that these could be incorporated into a rice crop simulation model such as CERES-RICE. Long-term soil processes also need to be validated.

As already mentioned in Section 15.5, biological N fixation is another area in which, surprisingly, there are few practical models available. Sheehy et al. (1987) did model some of the processes at the individual nodule level, but, to our knowledge, this has not been incorporated into a larger plant growth model. Some of the DSSAT legume crop models (e.g. CROPGRO) have a simple subroutine describing N fixation, but this has not been tested across a wide range of environments, and has never been published (G. Hoogenboom, University of Georgia, 2000, personal communication). Such a model would be of considerable use, for example, in determining 'best-bet' soil fertility technologies for different environ-mental conditions (taking into account temperature, water status and host plant growth), such as that currently being carried out in Nepal (Matthews, 2000). Understanding and predicting biological N fixation is likely to be of increasing importance with the growing concern with the over-use of artificial N fertilizers, particularly as N pollution has been highlighted as a major environmental problem of the next century (UNEP, 1999).

Soil acidity is one of the factors that is often blamed for a decline in crop yields in the tropics. A decrease in soil pH can be a result of: (i) the acidifying effect of N fertilizers; (ii) N fixation by legumes; and (iii) high degree of leaching. In Nepal, for example, farmers often complain of a deterioration or 'hardening' of their soil and a decline in yields due to the use of artificial fertilizers, and some are becoming reluctant to use them. With the exception of that of van Keulen (1995), few models explicitly take soil pH into account.

Salinity is another very important problem in some areas of the tropics, particularly where irrigation has been practised for long periods of time. Some progress has been made in enhancing models to take the effects of salinity into account (e.g. Asch *et al.*, 1997; Castrignano *et al.*, 1998), but these submodels have not yet been tested extensively for general use.

15.8 Summary

In this chapter, we have highlighted some of the key areas in which we feel that modelling has a useful role to play within natural resources systems approaches. This has necessarily been broader than our original intention to review the role of crop–soil models, but we have done this as we believe that there must be a shift in the thinking of crop–soil modellers towards: (i) making *people* more centre stage, and (ii) a more problem-solving approach.

On one level, this means thinking of the problems faced by ordinary people in developing countries, and constructing and applying their models to address, and contribute to solving, these problems. For this to be effective, modellers need to define clearly who the end-users of their models are, and to enter into dialogue with these people so that the final product is tailored to their needs. As part of this process, increased effort to disseminate the outputs of models, more than the models themselves, needs to be made.

On a second level, there is a need to consider people themselves as integral components in the systems being modelled. The SL framework currently being promoted by a number of international development organizations offers a good starting point from which to develop this methodology. It is hoped that this will eventually lead to the development of tools that practitioners could use to identify the real constraints to improved livelihoods in developing countries, so that future projects would be more realistically focused, thereby increasing their chances of impact.

Concluding Remarks 16

Robin Matthews and William Stephens

Institute of Water and Environment, Cranfield University, Silsoe, Bedfordshire MK45 4DT, UK

In this book, we have attempted to describe the major uses to which crop–soil simulation models have been put in recent years. In Chapter 1, we used Sinclair and Seligman's (1996) analogy relating the growth and development of crop simulation models to that of human beings. At the risk of pushing the analogy too far, this review has shown that, in many respects, crop–soil models have acquitted themselves well in their first employment, although it could be argued that they still have some way to go before they have 'security of tenure'.

It is clear that their greatest use so far has been by the research community, as models are primarily tools for organizing knowledge gained in experimentation. However, there is an urgent need to: (i) make the use of models in research more relevant to problems in the real world, and (ii) find effective means of dissemination of results from work using models to potential beneficiaries. In terms of decision support systems (DSSs) for practical use, the use of models has had major impacts in the areas of irrigation scheduling and pest management. However, particularly in pest management, this has not been as was originally intended – as soon as farmers have learned the optimal times to control pests, they have no further use for an operational DSS. Models are probably more useful as research tools to provide solutions to constraints which can then be developed into simple rules of thumb, rather than as operational decision support tools. Models also have a useful role to play as tools in education, both as aids to learning principles of crop and soil management, and also in helping students to develop a 'systems' way of thinking, to enable them to appreciate that their speciality is part of a larger system.

Various factors can prevent the use of models having any impact on target agricultural systems. These were categorized as: (i) inherent limitations of the models themselves for a particular use; (ii) constraints to the uptake and use of these models by target groups; and (iii) factors restricting the use of these models being translated into impact. Factors enhancing the likelihood of models making a significant impact include: the involvement of competent modellers working in multidisciplinary teams; the adoption of participatory approaches with end-users; a clear definition of problems being addressed; the demand for solutions from a target group; long-term commitment from funding sources; the application of models to variable environments where answers to problems are not obvious; and the need for quick answers.

In developing countries, crop–soil simulation models should be targeted in the near future at researchers, educationalists and agricultural consultants. It is unlikely that such models will provide a useful resource for extension personnel or farmers themselves in this time frame, although in the longer term, with increasing availability of computers and the skills to operate them, extension personnel may eventually find them appropriate. Nevertheless, the information generated by crop–soil simulation models can be useful for extension staff and farmers when incorporated into DSSs aimed at helping these groups choose the most appropriate techniques and land-use strategies for specific biophysical and socio-economic conditions. It is unlikely that such models would be incorporated directly into these DSSs, but they should be used wherever possible to develop the underlying knowledge base. The problem then becomes a general one of how to improve dissemination pathways between research and farmer. The process may be aided if the development of these DSSs involves the direct participation of end-users to ensure that the system is relevant to their needs. Information contained in, or generated by, the models needs to be transformed into other forms more appropriate to uptake by farmers (e.g. posters, booklets, radio, etc.). Participatory exercises involving modellers and farmers interacting together to evaluate farming strategies have been shown to greatly increase the likelihood of uptake and should therefore be encouraged.

In many ways, crop modelling as a research area is at a crossroads. Mathematical representations for most of the major crop processes such as photosynthesis, transpiration, respiration, N uptake, and so on, have now been developed. There is probably some scope for further refinement of some of these – for example, the underlying processes involved in allocation of assimilate to the various plant components are still not well understood (Hammer, 1998), and we have already mentioned some other areas in which crop–soil models can still be improved (Chapter 15). Nevertheless, existing models do have ways of handling these processes, and while a more detailed understanding may help to refine them, they are really just fine-tuning and are unlikely to contribute significantly to improving the accuracy and reliability of the models at the crop level. What then are the areas towards which crop simulation research is heading?

We would suggest that there are two opposite directions, both of which were discussed in Chapter 15. On the one hand, the rapidly expanding field of genomics means that links between information at the gene level and performance at the phenotype level need to be established, and methodologies developed to do this. The emerging fields of proteomics and 'metabolomics' will eventually provide the basic knowledge of how these links operate, which can then be incorporated into models. Such models will have the potential to contribute to improving the efficiency of crop improvement programmes worldwide by providing more efficient ways of identifying and evaluating desirable characteristics for specific plant breeding goals.

In the other direction, we see an area of research developing in which crop models are incorporated into higher-order systems such as the whole farm, catchment or region. At one level, linking crop growth models with other physical process models, such as those describing soil processes influencing gaseous emissions, for example, is a logical next step and is occurring to some extent already. However, there is a need to go further, particularly in integrating the outputs from models to improve our understanding of how changes in agricultural systems influence overall environmental impacts. As an example, in rice production systems, CO_2, temperature, methane (CH_4) emissions, and nitrous oxide (N_2O) emissions have all been studied as separate factors in detail, but there have been few studies in which these have been studied together as part of an agricultural system. Field drainage to reduce CH_4 emissions, for example, can increase N_2O emissions, so that the overall effect on global warming may be negligible. Such studies have a vital role to play in environmental research in developing countries, and can have a major impact by helping those countries meet their obligations under the United Nations Framework Convention on Climate Change.

At another level, the role of people in these systems also needs to be made explicit, so that the day-to-day decisions that they take to sustain and improve their livelihoods and the influence these have on their environment and natural resource base can be taken into account. In the rice example in the previous paragraph, models may well demonstrate the environmental benefits of changing agricultural techniques, but there will be no impact if resource-poor farmers do not recognize direct and immediate benefits to their livelihoods were they to adopt these techniques. In our view, the most challenging task is to make the virtual world inhabited by scientists and modellers resemble more closely the reality experienced by farmers. To bring these two worlds closer together, ways must be found of incorporating knowledge in such diverse fields as agronomy, agrometeorology, soil erosion, livestock, sociology, economics and policy making into workable and realistic whole-farm and community simulation models so that practitioners have confidence in the outputs and are willing to adapt them to their own situation.

As part of this process, the application of artificial intelligence techniques such as Bayesian belief networks, cellular automata and multi-agent systems simulation also need to be evaluated and adapted to natural resources systems. Such models have the potential to explore the likely outcomes of different policies on both the livelihoods of the people affected, and on the environment in which they live. This potential, however, will only be attained through the active cooperation between modellers and those they seek to serve. This requires recognition by end-users and beneficiaries that models are useful tools that can contribute to solving their problems, and by modellers that models alone are not enough to solve these problems.

References

Abawi, G.Y. (1993) Optimising harvest operations against weather risk. In: Penning de Vries, F.W.T., Teng, P. and Metselaar, K. (eds) *Systems Approaches for Agricultural Development*. Kluwer, Dordrecht, pp. 127–143.

Abrecht, D.G., Robinson, S.D., Henderson Sellers, B., McAleer, M. and Jakeman, A.J. (1996) TACT: a tactical decision aid using a CERES based wheat simulation model. *Ecological Modelling* 86(2–3), 241–244.

Ackoff, R.L. (1981) The art and science of mess management. *Interfaces* 11, 20–25.

Acosta-Gallegos, J.A. and White, J.W. (1995) Phenological plasticity as an adaptation by common bean to rainfed environments. *Crop Science* 35, 199 204.

Aggarwal, P.K. (1991) Estimation of the optimal duration of wheat crops in rice-wheat cropping systems by crop growth simulation. In: Penning de Vries, F.W.T., van Laar, H.H. and Kropff, M.J. (eds) *Simulation and Systems Analysis for Rice Production (SARP)*. Pudoc, Wageningen, pp. 3–10.

Aggarwal, P.K. (1993) Agro-ecological zoning using crop growth simulation models: characterisation of wheat environments in India. In: Penning de Vries, F.W.T., Teng, P. and Metselaar, K. (eds) *Systems Approaches for Agricultural Development*. Kluwer, Dordrecht, pp. 97–109.

Aggarwal, P.K. (1995) Uncertainties in crop, soil and weather inputs used in growth models: implications for simulated outputs and their applications. *Agricultural Systems* 48, 361–384.

Aggarwal, P.K. and Kalra, N. (1994) Analysing the limitations set by climatic factors, genotype, and water and nitrogen availability on productivity of wheat: II. Climatically potential yields and management strategies. *Field Crops Research* 38, 93–103.

Aggarwal, P.K. and Penning de Vries, F.W.T. (1989) Potential and water-limited yields in South-east Asia. *Agricultural Systems* 30, 49–69.

Aggarwal, P.K., Kalra, N., Bandyopadhyay, S.K. and Selvarajan, S. (1995) A systems approach to analyze production options for wheat in India. In: Bouma, J., Kuyvenhoven, A., Bouman, B.A.M., Luten, J.C. and Zandstra, H.G. (eds) *Eco-regional Approaches for Sustainable Land Use and Food Production. Systems Approaches for Sustainable Agricultural Development.* Kluwer, Dordrecht, pp. 167–186.

Aggarwal, P.K., Kropff, M.J., Matthews, R.B. and McLaren, C.G. (1996) Using simulation models to design new plant types and to analyse genotype by environment interactions in rice. In: Cooper, M. and Hammer, G.L. (eds) *Plant Adaptation and Crop Improvement.* CAB International, Wallingford, UK, pp. 403–418.

Aggarwal, P.K., Kropff, M.J., Cassman, K.G. and ten Berge, H.F.M. (1997) Simulating genotype strategies for increasing rice yield potential in irrigated, tropical environments. *Field Crops Research* 51, 5–17.

Aggarwal, P.K., Kalra, N., Kumar, S., Bandyopadhyay, S.K., Pathak, H., Vasisht, A.K. and Roetter, R.P. (2000a) Exploring land-use options for a sustainable increase in food grain production in Haryana: methodological framework. In: Roetter, R.P., van Keulen, H., Laborte, A.G., Hoanh, C.T. and van Laar, H.H. (eds) *Systems Research for Optimising Future Land Use in South and Southeast Asia.* International Rice Research Institute, Makati, The Philippines, pp. 57–68.

Aggarwal, P.K., Kalra, N., Kumar, S., Pathak, H., Bandyopadhyay, S.K., Vasisht, A.K., Roetter, R.P. and Hoanh, C.T. (2000b) Haryana State case study: trade-off between cere-al production and environmental impact. In: Roetter, R.P., van Keulen, H. and van Laar, H.H. (eds) *Synthesis of Methodology Development and Case Studies.* SysNet Research Paper Series No. 3. International Rice Research Institute, Makati, The Philippines, pp. 11–18.

Alexandrov, V. (1997) Vulnerability of agronomic systems in Bulgaria. *Climatic Change* 36, 135–149.

Alocilja, E.C. and Ritchie, J.T. (1993) Multicriteria optimisation for a sustainable agriculture. In: Penning de Vries, F.W.T., Teng, P. and Metselaar, K. (eds) *Systems Approaches for Agricultural Development.* Kluwer, Dordrecht, pp. 381–396.

Amien, I., Rejekingrum, P., Pramudia, A. and Susanti, E. (1996) Effects of interannual climate variability and climate change on rice yield in Java, Indonesia. *Water Air and Soil Pollution* 92, 29–39.

Anderson, J.R. and Hardaker, J.B. (1992) Efficacy and efficiency in agricultural research: a systems view. *Agricultural Systems* 40, 105–123.

Angus, J.F. (1989) Simulation models of water balance and growth of rainfed rice crops grown in sequence. ACIAR Monograph. Australian Council of International Agricultural Research, Canberra.

Angus, J.F. (1991) The evolution of methods for quantifying risk in water limited environments. In: Muchow, R.C. and Bellamy, J.A. (eds) *Climatic Risk in Crop Production – Models and Management for the Semi-arid Tropics and Sub-tropics.* CAB International, Wallingford, UK, pp. 39–53.

Arah, J.R.M. and Kirk, G.J.D. (2000) Modelling rice-plant-mediated methane emission. *Nutrient Cycling in Agroecosystems* 58, 221–230.

Asch, F., Dingkuhn, M., Wopereis,

M.C.S., Dörffling, K. and Miézan, K. (1997) A conceptual model for sodium uptake and distribution in irrigated rice. In: Kropff, M.J., Teng, P.S., Aggarwal, P.K., Bouma, J., Bouman, B.A.M., Jones, J.W. and van Laar, H.H. (eds) *Applications of Systems Approaches at the Field Level. Systems Approaches for Sustainable Agricultural Development.* Kluwer, Dordrecht, pp. 177–187.

Ashley, C. and Carney, D. (1999) *Sustainable Livelihoods: Lessons from Early Experience.* Department for International Development, London.

Bachelet, D. and Gay, C.A. (1993) The impacts of climate change on rice yield: a comparison of four model performances. *Ecological Modelling* 65, 71–93.

Bachelet, D., van Sickle, J. and Gay, C.A. (1993) The impacts of climate change on rice yield: evaluation of the efficacy of different modelling approaches. In: Penning de Vries, F.W.T., Teng, P.S. and Metselaar, K. (eds) *Systems Approaches for Agricultural Development.* Kluwer, Dordrecht, pp. 145–174.

Bailey, E. and Boisvert, R.N. (1989) A comparison of risk efficiency criteria in evaluating groundnut performance in drought-prone areas. *Australian Journal of Agricultural Economics* 33, 153–169.

Baker, J.T., Allen, L.H., Boote, K.J., Jones, P. and Jones, J.W. (1990) Developmental responses of rice to photoperiod and carbon dioxide concentration. *Agricultural Forest Meteorology* 50, 201–210.

Baker, M.J. (1996) Use and abuse of crop simulation models – Forward. *Agronomy Journal* 88, 689.

Baker, R.J. (1988) Differential response to environmental stress. In: Weir, B.S., Eisen, E.J., Goodman, M.M. and Namkoong, G. (eds) *Proceedings of the Second International Conference on Quantitative Genetics.* Sinauer Associates, Sunderland, Massachusetts, pp. 492–504.

Bannayan, M. and Crout, N.M.J. (1999) A stochastic modelling approach for real-time forecasting of winter wheat yield. *Field Crops Research* 62, 85–95.

Barbour, J.C. and Bridges, D.C. (1995) A model of competition for light between peanut (*Arachis hypogaea*) and broadleaf weeds. *Weed Science* 43, 247–257.

Barlow, N.D. (1998) Models in biological control: a field guide. In: Hawkins, B.A. and Cornell, H.B. (eds) *Theoretical Approaches to Biological Control.* Cambridge University Press, Cambridge, pp. 43–68.

Barnum, H.N. and Squire, L. (1979) A model of an agricultural household: theory and evidence. Occasional Paper No. 27. World Bank, Washington, DC.

Barreteau, O., Bousequet, F. and Attonaty, J.-M. (2001) Role-playing games for opening the black box of multi-agent systems: method and lessons of its application to Senegal River Valley irrigated systems. *Journal of Artificial Societies and Social Simulation* 4: http://www.soc.surrey.ac.uk/JASSS/4/2/5.html

Bastiaans, L., Kropff, M.J., Kempuchetty, N., Rajan, A. and Migo, T.R. (1997) Can simulation models help design rice cultivars that are more competitive against weeds? *Field Crops Research* 51, 101–111.

Batchelor, W.D. (1997) PLANTMOD 2.1, exploring the physiology of plant communities. *Field Crops Research* 54, 87–88.

Batchelor, W.D., Jones, J.W., Boote, K.J. and Pinnschmidt, H.O. (1993) Extending the use of crop models to study pest damage. *Transactions of* the American Society of Agricultural Engineering 36, 551–558.

Becker, G.S. (1965) A theory of the allocation of time. *Economic Journal* 75, 493–517.

Beinroth, F.H., Jones, J.W., Knapp, E.B., Papajorgji, P. and Luyten, J. (1998) Evaluation of land resources using crop models and a GIS. In: Tsuji, G.Y., Hoogenboom, G. and Thornton, P.K. (eds) *Understanding Options for Agricultural Production. Systems Approaches for Sustainable Agricultural Development.* Kluwer, Dordrecht, pp. 293–311.

Bell, M.A. and Fischer, R.A. (1994) Using yield prediction models to assess wheat gains: a case study for wheat. *Field Crops Research* 36, 161–166.

Bernet, T., Ortiz, O., Estrada, R.D., Quiroz, R. and Swinton, S.M. (2001) Tailoring agricultural extension to different production contexts: a user-friendly farm-household model to improve decision-making for participatory research. *Agricultural Systems* 69, 183–198.

Bidinger, F.R., Hammer, G.L. and Muchow, R.C. (1996) The physiological basis of genotype by environment interaction in crop adaptation. In: Cooper, M. and Hammer, G.L. (eds) *Plant Adaptation and Crop Improvement.* CAB International, Wallingford, UK, pp. 329–347.

Boote, K.J., Batchelor, W.D., Jones, J.W., Pinnschmidt, H. and Bourgeois, G. (1993) Pest damage relations at the field level. In: Penning de Vries, F.W.T., Teng, P. and Metselaar, K. (eds) *Systems Approaches for Agricultural Development.* Kluwer, Dordrecht, pp. 277–296.

Boote, K.J. and Jones, J.W. (1988) Applications of, and limitations to, crop growth simulation models to fit crops and cropping systems to semi-arid environments. In: Bidinger, F.R. and Johansen, C. (eds) *Drought Research Priorities for the Dryland Tropics.* ICRISAT, Patancheru, India, pp. 63–65.

Boote, K.J. and Tollenaar, M. (1994) *Modelling Yield Potential, Physiology, Determination of Crop Yield.* American Society of Agronomy, Madison, Wisconsin, pp. 553–565.

Boote, K.J., Jones, J.W. and Singh, P. (1991) Modeling growth and yield of groundnut: state of the art. In: *Groundnut: a Global Perspective. Proceedings of an International Workshop,* 25–29 November 1991. ICRISAT, Patancheru, India, pp. 331–343.

Boote, K.J., Jones, J.W. and Pickering, N.B. (1996) Potential uses and limitations of crop models. *Agronomy Journal* 88, 704–716.

Bouma, J., Kuyvenhoven, A., Bouman, B.A.M., Luten, J.C. and Zandstra, H.G. (eds) (1995) *Eco-regional Approaches for Sustainable Land Use and Food Production. Systems Approaches for Sustainable Agricultural Development,* 4. Kluwer, Dordrecht.

Bouman, B.A.M., van Keulen, H. and Rabbinge, R. (1996) The 'School of de Wit' crop growth simulation models: a pedigree and historical overview. *Agricultural Systems* 52, 171–198.

Bouman, B.A.M., Schipper, R.A., Nieuwenhuyse, A., Hengsdijk, H. and Jansen, H.G.P. (1998) Quantifying economic and biophysical sustainability trade-offs in land-use exploration at the regional level: a case study for the Northern Atlantic Zone of Costa Rica. *Ecological Modelling* 114, 95–109.

Bouman, B.A.M., Jansen, H.G.P., Schipper, R.A., Nieuwenhuyse, A.,

Hengsdijk, H. and Bouma, J. (1999) A framework for integrated bio-physical and economic land use analysis at different scales. *Agricultural Ecosystems and Environment* 75, 55–73.

Bouwman, A.F. (1991) Agronomic aspects of wetland rice cultivation and associated methane emissions. *Biogeochemistry* 15, 65–88.

Bowen, W.T. and Baethgen, W.E. (1998) Simulation as a tool for improving nitrogen management. In: Tsuji, G.Y., Hoogenboom, G. and Thornton, P.K. (eds) *Understanding Options for Agricultural Production. Systems Approaches for Sustainable Agricultural Development.* Kluwer, Dordrecht, pp. 189–204.

Bowen, W.T. and Papajorgji, P. (1992) DSSAT estimated wheat productivity following late-season nitrogen application in Albania. *Agrotechnology Transfer* 16, 9–12.

Bowen, W.T., Jones, J.W., Carsky, R.J. and Quintana, J.O. (1992) Evaluation of the nitrogen submodel of CERES-Maize following legume green manure incorporation. *Agronomy Journal* 85, 153–159.

Bowen, W.T., Thornton, P.K. and Hoogenboom, G. (1998) The simulation of cropping sequences using DSSAT. In: Tsuji, G.Y., Hoogenboom, G. and Thornton, P.K. (eds) *Understanding Options for Agricultural Production. Systems Approaches for Sustainable Agricultural Development.* Kluwer, Dordrecht, pp. 313–327.

Bradley, R.G. and Crout, N.M.J. (1994) *The PARCH Model for Predicting Arable Resource Capture in Hostile Environments: Users' Guide.* Tropical Crops Research Unit, University of Nottingham, Sutton Bonington, UK.

Breman, H. (1990) Integrating crops and livestock in southern Mali: rural development or environmental degradation? In: Rabbinge, R., Goudriaan, J., van Keulen, H., Penning de Vries, F.W.T. and van Laar, H.H. (eds) *Theoretical Production Ecology: Reflection and Prospects.* Simulation Monographs, Pudoc, Wageningen, pp. 277–301.

Brooks, R. (1990) Elephants don't play chess. *Robotics and Autonomous Systems* 6, 3–15.

Bruentrup, M., Lamers, J.P.A. and Herrmann, L. (1997) Modelling the long-term effects of crop residue management for sustainable farming systems. In: Teng, P.S., Kropff, M.J., ten Berge, H.F.M., Dent, J.B., Lansigan, F.P. and van Laar, H.H. (eds) *Applications of Systems Approaches at the Farm and Regional Levels. Systems Approaches for Sustainable Agricultural Development.* Kluwer, Dordrecht, pp. 53–64.

Buan, R.D., Maglinao, A.R., Evangelista, P.P. and Pajeulas, B.G. (1996) Vulnerability of rice and corn to climate change in the Philippines. *Water, Air and Soil Pollution* 92, 41–51.

Burrough, P.A. (1989a) Matching spatial databases and quantitative models in land resource assessment. *Soil Use and Management* 5, 3–8.

Burrough, P.A. (1989b) Modelling land qualities in space and time: the role of geographical information systems. In: Bouma, J. and Bregt, A.K. (eds) *Land Qualities in Space and Time. Proceedings of the ISSS Symposium.* Pudoc, Wageningen, pp. 45–60.

Burton, M.A. (1989) Experiences with the Irrigation Management Game. *Irrigation and Drainage Systems* 3, 217–228.

Cabelguenne, M. (1996) Tactical irrigation management using real time EPIC-phase model and weather forecast: experiment on maize. In: *Irrigation Scheduling from Theory to Practice (Water Reports)*. FAO-ICID-CIID, Rome, pp. 185–193.

Calvero, S.B. and Teng, P.S. (1997) Use of simulation models to optimise fungicide use for managing tropical rice blast disease. In: Kropff, M.J., Teng, P.S., Aggarwal, P.K., Bouma, J., Bouman, B.A.M., Jones, J.W. and van Laar, H.H. (eds) *Applications of Systems Approaches at the Field Level. Systems Approaches for Sustainable Agricultural Development*. Kluwer, Dordrecht, pp. 305–320.

Carney, D. (1998) Implementing the sustainable rural livelihoods approach. In: *Sustainable Rural Livelihoods – what contribution can we make?* Department for International Development, London, pp. 3–23.

Castrignano, A., Katerji, N., Karam, F., Mastrorilli, M. and Hamdy, A. (1998) A modified version of CERES-Maize model for predicting crop response to salinity stress. *Ecological Modelling* 111, 107–120.

Caton, B.P., Foin, T.C. and Hill, J.E. (1999) A plant growth model for integrated weed management in direct-seeded rice. III. Interspecific competition for light. *Field Crops Research* 63(1), 47–61.

Chapman, S.C. and Bareto, H.J. (1996) Using simulation models and spatial databases to improve the efficiency of plant breeding programs. In: Cooper, M. and Hammer, G.L. (eds) *Plant Adaptation and Crop Improvement*. CAB International, Wallingford, UK, pp. 563–587.

Chapman, S.C., Cooper, M., Butler, D.G. and Hammer, G.L. (2000a) Genotype by environment interactions affecting grain sorghum. III. Temporal sequences and spatial patterns in the target population of environments. *Australian Journal of Agricultural Research* 51, 223–33.

Chapman, S.C., Cooper, M., Hammer, G.L. and Butler, D.G. (2000b) Genotype by environment interactions affecting grain sorghum. II. Frequencies of different seasonal patterns of drought stress are related to location effects on hybrid yields. *Australian Journal of Agricultural Research* 51, 209–221.

Chapman, S., Cooper, M., Podlich, D. and Hammer, G. (2002) Evaluating plant breeding strategies by simulating gene action and environment effects to predict phenotypes for dryland adaptation. *Agronomy Journal* (in press).

Checkland, P. (1981) *Systems Thinking, Systems Practice*. John Wiley & Sons, Chichester, UK.

Chou, T.Y. and Chen, H.Y. (1995) The application of a decision-support system for agricultural land management. *Journal of Agriculture and Forestry* 44, 75–89.

Clegg, C., Axtell, C., Damodaran, L., Farbey, B., Lloyd-Jones, R., Nicholls, J., Sell, R., Tomlinson, C., Ainger, A. and Stewart, T. (1996) *The Performance of Information Technology and the Role of Human and Organizational Factors*. Institute of Work Psychology, University of Sheffield, Sheffield.

Clewett, J.F., Howden, S.M., McKeon, G.M. and Rose, C.W. (1991) Optimising farm dam irrigation in response to climate risk. In: Muchow, R.C. and Bellamy, J.A. (eds) *Climatic Risk in Crop Production – Models and Management for the Semi-arid Tropics and Sub-tropics*. CAB International, Wallingford, UK, pp. 307–328.

CLUES (1996a) Challenging students with simulation models. *CTI Newsletter* 16.

CLUES (1996b) Learning by simulation. *CTI Newsletter* 16.

CLUES (1998) BAITA: new cross sector

association for IT in Agriculture. *CTI Newsletter* 22(6).

Cock, J.H., Franklin, D., Sandoval, G. and Juri, P. (1979) The ideal cassava plant for maximum yield. *Crop Science* 19, 271–279.

Cohen, S.J. (1990) Bringing the global warming issue closer to home: the challenge of regional impact studies. *Bulletin of the American Meteorological Society* 71, 520–526.

Coleman, J.N., Kinnimet, D.J., Burns, F.P., Butler, T.J. and Koelmans, A.M. (1998) Effectiveness of computer-aided learning as a direct replacement for lecturing in degree-level electronics. *IEEE Transactions on Education* 41, 177–184.

Collinson, M.P. and Tollens, E. (1994) The impact of the international agricultural centres: measurement, quantification and interpretation. *Experimental Agriculture* 30, 395–419.

Comstock, R.E. and Moll, R.H. (1963) Genotype–environment interactions. In: Hanson, W.D. and Robinson, H.F. (eds) *Statistical Genetics and Plant Breeding.* NAS-NRC, Washington, DC, pp. 164–196.

Cooper, M. and Fox, P.N. (1996) Environmental characterisation based on probe and reference genotypes. In: Cooper, M. and Hammer, G.L. (eds) *Plant Adaptation and Crop Improvement.* CAB International, Wallingford, UK, pp. 529–547.

Cooper, M. and Hammer, G.L. (1996) Synthesis of strategies for crop improvement. In: Cooper, M. and Hammer, G.L. (eds) *Plant Adaptation and Crop Improvement.* CAB International, Wallingford, UK, pp. 591–623.

Cooper, M., Podlich, D.W., Jensen, N.M., Chapman, S.C. and Hammer, G.L. (1999) Modelling plant breeding programs. *Trends in Agronomy* 2, 33–64.

Cox, P.G. (1996) Some issues in the design of agricultural decision support systems. *Agricultural Systems* 52, 355–381.

Cure, J.D. and Acock, B. (1986) Crop responses to carbon dioxide doubling: a literature survey. *Agricultural and Forest Meteorology* 38, 127–145.

de Jager, A., Nandwa, S.M. and Okoth, P.F. (1998) Monitoring nutrient flows and economic performance in African farming systems (NUTMON). I. Concepts and methodologies. *Agricultural Ecosystems and Environment* 71, 37–48.

de Jager, J.M., Hoffman, J.E., van Feden, F., Pretorius, J., Marais, J., Erasmus, J.F., Cowley, B.S. and Mottram, R. (1983) Preliminary validation of the PUTU maize crop growth model in different parts of South Africa. *Crop Production* 12, 3–6.

de Moed, G.H., van der Werf, W. and Smits, P.H. (1990) Modelling the epizootiology of *Spodoptera exigua* nuclear polyhedrosis virus in a spatially distributed population of *Spodoptera exigua* in greenhouse Chrysanthemums. *SROP/WPRS Bulletin* XIII, 135–141.

de Wit, C.T. (1965) *Photosynthesis of Leaf Canopies.* Pudoc, Wageningen.

Defeng, Z. and Shaokai, M. (1995) Rice production in China under current and future climates. In: Matthews, R.B., Kropff, M.J., Bachelet, D. and van Laar, H.H. (eds) *Modelling the Impact of Climate Change on Rice Production in Asia.* IRRI/CAB International, Wallingford, UK, pp. 217–235.

DeLacy, I.H., Basford, K.E., Cooper, M., Bull, J.K. and McLaren, C.G. (1996) Analysis of multi-environment trials – an historical perspective. In: Cooper, M. and Hammer, G.L. (eds) *Plant Adaptation and Crop*

Improvement. CAB International, Wallingford, UK, pp. 39–124.

Dempster, J.I.M., Marsden, S.K. and Smout, I.K. (1989) Computer simulation games for training in irrigation management. *Irrigation and Drainage Systems* 3, 265–280.

Dent, J.B. (1993) Potential for systems simulation in farming systems research. In: Penning de Vries, F.W.T., Teng, P. and Metselaar, K. (eds) *Systems Approaches for Agricultural Development.* Kluwer, Dordrecht, pp. 325–339.

Dent, J.B. and Edwards-Jones, G. (1991) The context of modelling in the future: the changing nature of R&D funding in future. *Aspects of Applied Biology* 26, 183–193.

Dent, J.B. and Thornton, P.K. (1988) The role of biological simulation models in farming systems research. *Agricultural Administration and Extension* 29, 111–122.

Deunier, J.M., Leroy, P. and Peyremorte, P. (1996) Tools for improving management of irrigated agricultural crop systems. In: *Irrigation Scheduling from Theory to Practice (Water Reports).* FAO-ICID-CIID, Rome, pp. 39–49.

Dillon, C.R., Mjelde, J.W. and McCarl, B.A. (1989) Biophysical simulation in support of crop production decisions: a case study in the Blacklands region of Texas. *Journal of Agricultural Economics* 21, 73–85.

Dingkuhn, M. and Sow, A. (1997) Potential yield of irrigated rice in African arid environments. In: Kropff, M.J., Teng, P.S., Aggarwal, P.K., Bouma, J., Bouman, B.A.M., Jones, J.W. and van Laar, H.H. (eds) *Applications of Systems Approaches at the Field Level. Systems Approaches for Sustainable Agricultural Development.* Kluwer, Dordrecht, pp. 79–99.

Dingkuhn, M., Penning de Vries, F.W.T., De Datta, S.K. and van Laar, H.H. (1991) Concepts for a new plant type for direct seed flooded tropical rice. In: *Direct Seeded Flooded Rice in the Tropics: Selected Papers from the International Rice Research Conference.* International Rice Research Institute, Los Banos, The Philippines, pp. 17–38.

Dingkuhn, M., Jones, M.P., Fofana, B. and Sow, A. (1997) New high yielding, weed competitive rice plant types drawing from *O. sativa* and *O. glaberrima* genepools. In: Kropff, M.J., Teng, P.S., Aggarwal, P.K., Bouma, J., Bouman, B.A.M., Jones, J.W. and van Laar, H.H. (eds) *Applications of Systems Approaches at the Field Level. Systems Approaches for Sustainable Agricultural Development.* Kluwer, Dordrecht, pp. 37–52.

Djurle, J. (1988) Experience and results from the use of EPIPRE in Sweden. *SROP Bulletin.* In: *Pest and Disease Models in Forecasting Crop Loss Appraisal and Decision Supported Crop Protection Systems* 11, 94–95.

Donald, C.M. (1968) The breeding of crop ideotypes. *Euphytica* 17, 385–403.

Donald, C.M. and Hamblin, J. (1983) The convergent evolution of annual seed crops in agriculture. *Advances in Agronomy* 36, 97–143.

Doorenbos, J. and Kassam, A.H. (1979) Yield Response to Water. Irrigation and Drainage Paper No 33, Food and Agricultural Organisation, Rome.

Doyle, C.J. (1991) Mathematical models in weed management. *Crop Protection* 10, 432–444.

Doyle, C.J. (1997) A review of the use of models of weed control in integrated crop protection. *Agricultural*

Ecosystems and Environment 64, 165–172.

du Pisani, A.L. (1987) The CERES-Maize model as a potential tool for drought assessment in South Africa. *Water SA* 13, 159–164.

Dua, A.B., Penning de Vries, F.W.T. and Seshu, D.V. (1990) Simulation to support evaluation of the production potential of rice varieties in tropical climates. *Transactions of the ASAE* 33, 1185–1194.

Dudley, N.J. and Hearn, A.B. (1993) Systems modelling to integrate river valley water supply and irrigation decision-making under uncertainty. *Agricultural Systems* 42, 3–23.

Duncan, W.G., McCloud, D.E., McGraw, R.L. and Boote, K.J. (1978) Physiological aspects of peanut yield improvement. *Crop Science* 18, 1015–1020.

Edward, N.S. (1996) Evaluation of computer based laboratory simulation. *Computers and Education* 26(1), 125–130.

Edwards, W.M. (1997) Teaching farm financial management by interactive simulation. *Agricultural Finance Review* 57(7), 81–91.

Edwards-Jones, G., Dent, J.B., Morgan, O. and McGregor, M.J. (1998) Incorporating farm household decision-making within whole farm models. In: Tsuji, G.Y., Hoogenboom, G. and Thornton, P.K. (eds) *Understanding Options for Agricultural Production.* Kluwer, Dordrecht, pp. 347–365.

Elwell, D.L., Curry, R.B. and Keener, M.E. (1987) Determination of potential yield-limiting factors of soybeans using SOYMOD/OARDC. *Agricultural Systems* 24, 221–242.

Falloon, P.D., Smith, J.U. and Smith, P. (1999) A review of decision support systems for fertiliser application and manure management. *Acta Agronomica Hungarica* 47, 227–236.

FAO (1976) *A Framework for Land Evaluation.* FAO Soils Bulletin, No. 32. Food and Agriculture Organisation, Rome.

Field, T.R.O. and Hunt, L.A. (1974) The use of simulation techniques in the analysis of seasonal changes in the productivity of alfalfa (*Medicago sativa* L.) stands. In: *Proceedings of the International Grasslands Congress XII,* Moscow, pp. 108–120.

Finlay, K.W. and Wilkinson, G.N. (1963) The analysis of adaptation in a plant breeding programme. *Australian Journal of Agricultural Research* 14, 742–754.

Finlay, P. (1989) *Introducing Decision Support Systems.* NCC Blackwell, Oxford.

Forrer, H.R. (1988) Experience and current status of EPIPRE in Switzerland. *SROP Bulletin.* In: *Pest and Disease Models in Forecasting Crop Loss Appraisal and Decision Supported Crop Protection Systems* 11, 91–93.

Forsyth, D.R. and Archer, C.R. (1997) Technologically assisted instruction and student mastery, motivation, and matriculation. *Teaching of Psychology* 24, 207–212.

Freebairn, D.M., Littleboy, M., Smith, G.D. and Coughlan, K.J. (1991) Optimising soil surface management in response to climate risk. In: Muchow, R.C. and Bellamy, J.A. (eds) *Climatic Risk in Crop Production – Models and Management for the Semi-arid Tropics and Sub-tropics.* CAB International, Wallingford, UK, pp. 283–305.

Fry, G.J. (1996) *PARCH Uptake and Impact Evaluation.* Natural Resources Institute, Chatham Maritime, UK.

Fujisaka, S. (1993) A case of farmer adaptation and adoption of contour hedgerows for soil conservation. *Experimental Agriculture* 29, 97–105.

Fukai, S. and Hammer, G.L. (1987) A simulation model of the growth of the cassava crop and its use to estimate cassava productivity in Northern Australia. *Agricultural Systems* 23, 237–257.

Garnett, E.R. and Khandekar, M.L. (1992) The impact of large-scale atmospheric circulations and anomalies on Indian monsoon droughts and floods and on world grain yields – a statistical analysis. *Agricultural and Forest Meteorology* 61, 113–128.

Gaunt, J.L. and Arah, J.R.M. (2001) *Modelling Soil Organic Matter Transformations and Nitrogen Availability in Periodically Flooded Soils.* IACR Rothamsted, Harpenden.

Gijsman, A.J., Oberson, A., Tiessen, H. and Friesen, D.K. (1996) Limited applicability of CENTURY model to highly weathered tropical soils. *Agronomy Journal* 88, 894–903.

Gijsman, A.J., Hoogenboom, G. and Parton, W.J. (1999) Linking DSSAT and CENTURY for improved simulation of smallholder agricultural systems. In: Donatelli, M., Stockle, C., Villalobos, F. and Mir, J.M.V. (eds), *Proceedings of the International Symposium on Modelling Cropping Systems,* 21–23 June 1999, Lleida, Spain. University of Lleida, Lleida, Spain, pp. 189–190.

Goldsworthy, P. and Penning de Vries, F.W.T. (1994) *Opportunities, Use, and Transfer of Systems Research Methods in Agriculture to Developing Countries. Systems Approaches for Sustainable Agricultural Development,* 3. Kluwer, Dordrecht.

Goodwin, P. and Wright, G. (1998) *Decision Analysis for Management Judgement.* John Wiley & Sons, Chichester, UK.

Graf, B. and Hill, J.E. (1992) Modelling the competition for light and nitrogen between rice and *Echinochloa crus-galli. Agricultural Systems* 40, 345–359.

Greer, J.E., Falk, S., Greer, K.J. and Bentham, M. (1995) Explaining and justifying recommendations in an agricultural decision support system. *Computers and Electronics in Agriculture* 11, 195–214.

Gundry, S. (1994) An integrated model of the food system in a region in Zimbabwe. Institute of Ecology and Resource Management, University of Edinburgh, Edinburgh.

Hamilton, W.D., Woodruff, D.R. and Jamieson, A.M. (1991) Role of computer-based decision aids in farm decision-making and in agricultural extension. In: Muchow, R.C. and Bellamy, J.A. (eds) *Climatic Risk in Crop Production – Models and Management for the Semi-arid Tropics and Subtropics.* CAB International, Wallingford, UK, pp. 411–423.

Hammer, G.L. (1998) Crop modelling: current status and opportunities to advance. *Acta Horticulturae* 456, 27–36.

Hammer, G.L. and Muchow, R.C. (1991) Quantifying climatic risk to sorghum in Australia's semi-arid tropics and subtropics: model development and simulation. In: Muchow, R.C. and Bellamy, J.A. (eds) *Climatic Risk in Crop Production: Models and Management for the Semi-arid Tropics and Subtropics.* CAB International, Wallingford, UK, pp. 205–232.

Hammer, G.L. and Muchow, R.C. (1992) The use of simulation modelling in decision-making in sorghum production. In: Foale, M.A., Henzell, R.G. and Vance, P.N. (eds) *Proceedings of the Second Australian Sorghum Conference.* Australian Institute of

Agricultural Science, Melbourne, pp. 146–158.

Hammer, G.L. and Vanderlip, R.L. (1989) Genotype-by-environment interaction in grain sorghum. III. Modelling the impact in field environments. *Crop Science* 29, 385–391.

Hammer, G.L., Butler, D.G., Muchow, R.C. and Meinke, H. (1996a) Integrating physiological understanding and plant breeding via crop modelling and optimisation. In: Cooper, M. and Hammer, G.L. (eds) *Plant Adaptation and Crop Improvement.* CAB International, Wallingford, UK, pp. 419–441.

Hammer, G.L., Holzworth, D.P. and Stone, R. (1996b) The value of skill in seasonal climate forecasting to wheat crop management in a region with high climatic variability. *Australian Journal of Agricultural Research* 47, 717–737.

Hansen, J., Fung, I., Lacis, A., Rind, D., Lebedeff, S., Ruedy, R. and Russell, G. (1988) Global climate changes as forecast by the Goddard Institute for Space Studies three-dimensional model. *Journal of Geophysical Research* 93, 9341–9364.

Hansen, J.W. (1995) A systems approach to characterizing farm sustainability PhD Thesis, University of Florida, Gainesville, Florida.

Hansen, J.W. and Jones, J.W. (1996) A systems framework for characterising farm sustainability. *Agricultural Systems* 51(2), 185–201.

Hansen, J.W., Beinroth, F.H. and Jones, J.W. (1998) Systems-based land-use evaluation at the South Coast of Puerto Rico. *Applied Engineering in Agriculture* 14, 191–200.

Hansen, J.W., Knapp, E.B. and Jones, J.W. (1997) Determinants of sustainability of a Colombian hillside farm. *Experimental Agriculture* 33, 425–448.

Hasegawa, H., Labavitch, J.M., McGuire, A.M., Bryant, D.C. and Denison, R.F. (1999) Testing CERES model predictions of N release from legume cover crop residue. *Field Crops Research* 63, 255–267.

Heong, K.L. (1990) *Crop Loss Assessments in Rice.* IRRI, Manila.

Herath, H.M.G., Hardaker, J.B. and Anderson, J.R. (1982) Choices of varieties by Sri Lankan rice farmers: comparing alternative decision models. *American Journal of Agricultural Economics* 64, 87–93.

Hess, T. and Stephens, W. (1998) The role of modelling in renewable natural resources systems research. Cranfield University for the ODA, Silsoe, UK.

Hess, T.M. (1990) Practical experiences of operating a farm irrigation scheduling service in England. *Acta Horticulturae* 278, 871–878.

Hess, T.M. (1996) A microcomputer scheduling program for supplementary irrigation. *Computers and Electronics in Agriculture* 15, 233–243.

Higgins, A.J., Muchow, R.C., Rudd, A.V. and Ford, A.W. (1998) Optimising harvest date in sugar production: a case study for the Mossman mill region in Australia. I. Development of operations research model and solution. *Field Crops Research* 57, 153–162.

Hilhorst, R. and Manders, R. (1995) Critical success and fail factors for implementation of knowledge based systems at the farm level. In: ten Cate, A.J.U., Clouaire, R.M., Dijkhuizen, A. and Lockhorst, C. (eds) *2nd IFAC/IFIP/EurAg Eng Workshop on Artificial Intelligence in Agriculture IFAC.* Elsevier, Amsterdam, pp. 297–300.

Hoogenboom, G., Jones, J.W. and White, J.W. (1988) Use of models in studies of drought. In: White, J.W., Hoogenboom, G., Ibarra, F.

and Singh, S.P. (eds) *Research on Drought Tolerance in Common Bean*. CIAT, Cali, pp. 192–230.

Hoogenboom, G., White, J.W., Acosta-Gallegos, J., Gaudiel, R.G., Myers, J.R. and Silbernagel, M.J. (1997) Evaluation of a crop simulation model that incorporates gene action. *Agronomy Journal* 89, 613–620.

Hoogenboom, G., Gadgil, S., Singh, P., Reddy, T.Y., Rao, P.R.S. and Murthy, S.K. (2001) Peanut yield variability in rainfed production systems of the semi-arid tropics. In: *Proceedings – Third International Symposium on Systems Approaches for Agricultural Development, Lima, 8–10 November 1999*. International Potato Centre (CIP), Lima.

Hook, J.E. (1994) Using crop models to plan water withdrawals for irrigation in dry years. *Agricultural Systems* 45, 271–289.

Horie, T. (1988) The effect of climatic variations on agriculture in Japan. 5: the effects on rice yields in Hokkaido. In: Parry, M.L., Carter, T.R. and Konijn, N.T. (eds) *The Impact of Climate Variations on Agriculture. Volume 1: Assessments in Cool Temperate and Cold Regions*. Kluwer, Dordrecht, pp. 809–826.

Horie, T., Yajima, M. and Nakagawa, H. (1992) Yield forecasting. *Agricultural Systems* 40, 211–236.

Houghton, J.T., Callandar, B.A. and Varney, S.K. (1992) *Climate Change 1992. The Supplementary Report to the IPCC Scientific Assessment*. Cambridge University Press, Cambridge.

Hsin-i, W., Childress, W.M., Yang, L., Spence, R.D. and Ren, J. (1996) An integrated simulation model for a semi-arid agroecosystem in the Loess Plateau of northwestern China. *Agricultural Systems* 52, 83–111.

Hundal, S.S., Prabhjyot, K. and Kaur, P. (1999) Evaluation of agronomic practices for rice using computer simulation model, CERES-Rice. *Oryza* 36, 63–65.

Hunt, L.A. (1993) Designing improved plant types: a breeder's viewpoint. In: Penning de Vries, F.W.T., Teng, P. and Metselaar, K. (eds) *Systems Approaches for Agricultural Development*. Kluwer, Dordrecht, pp. 3–17.

IPCC (1990) Introduction. In: Houghton, J.T., Jenkins, G.J. and Ephraums, J.J. (eds) *Climate Change: the IPCC Scientific Assessment*. Cambridge University Press, Cambridge.

IPCC (1996) XII. Summary for policymakers. In: Houghton, J.T., Meira-Filho, L.G., Chancellor, B.A., Kattenberg, A. and Maskell, K. (eds) *Climate Change 1995: The Scientific Basis of Climate Change*. Cambridge University Press, Cambridge.

IRRI (1989) *IRRI Towards 2000 and Beyond*. International Rice Research Institute, Los Baños, The Philippines.

IRRI (1993) *IRRI Rice Almanac 1993–1995*. International Rice Research Institute, Los Baños, The Philippines.

Ison, R.L. (1993) Soft systems: a non-computer view of decision support. In: Stuth, J.W. and Lyons, B.G. (eds) *Decision Support Systems for the Management of Grazing Lands. Man and the Biosphere*. UNESCO, Paris, pp. 83–122.

Itier, B. (1996) Applicability and limitations of irrigation scheduling methods techniques. In: *Irrigation Scheduling from Theory to Practice (Water Reports)*. FAO-ICID-CIID, Rome, pp. 19–32.

Jagtap, S.S., Abamu, F.J. and Kling, J.G. (1999) Long-term assessment of nitrogen and variety technologies on attainable maize yields in Nigeria using CERES-Maize. *Agricultural Systems* 60, 77–86.

Jansen, D.M. (1990) Potential rice yields in future weather conditions in different parts of Asia. *Netherlands Journal of Agricultural Science* 38, 661–680.

Jansen, D.M. and Schipper, R.A. (1995) A static, descriptive approach to quantify land use systems. *Netherlands Journal of Agricultural Science* 43, 31–46.

Jansen, H.G.P. (2001) A decade of interdisciplinary land use research in Costa Rica by the Research Program on Sustainability in Agriculture (REPOSA): achievements and lessons. In: *Proceedings – Third International Symposium on Systems Approaches for Agricultural Development, Lima, 8–10* November 1999. International Potato Centre (CIP), Lima.

Jeger, M.J. (1997) Approaches to integrated crop protection in university education and training. *Agricultural Ecosystems and Environment* 64, 173–179.

Jensen, N.F. (1975) Breeding strategies for winter wheat improvement. *Proceedings of the Winter Wheat Conference*, Zagreb, pp. 31–45.

Jones, C. (1983) The mobilization of women's labor for cash crop production: a game theoretical approach. *American Journal of Agricultural Economics* 67, 1049–1054.

Jones, C.A. and Kiniry, J.R. (1986) *CERES-Maize: A Simulation Model of Maize Growth and Development*. Texas A&M University Press, College Station, Texas.

Jones, C.A. and O'Toole, J.C. (1987) Application of crop production models in agroecological characterisation: simulation models for specific crops. In: Bunting, A.H.

(ed.) *Agricultural Environments – Characterisation, Classification, and Mapping*. CAB International, Wallingford, UK, pp. 199–209.

Jones, J.W. and Zur, B. (1984) Simulation of possible adaptive mechanisms in crops subjected to water stress. *Irrigation Science* 5, 251–264.

Jones, J.W., Bowen, W.T., Boggess, W.G. and Ritchie, J.T. (1993) Decision support systems for sustainable agriculture. In: Ragland, J. and Lal, R. (eds) *Technologies for Sustainable Agriculture in the Tropics*. American Society of Agronomy, Madison, Wisconsin, pp. 123–138.

Jones, J.W., Thornton, P.K. and Hansen, J.W. (1997) Opportunities for systems approaches at the farm scale. In: Teng, P.S., Kropff, M.J., ten Berge, H.F.M., Dent, J.B., Lansigan, F.P. and van Laar, H.H. (eds) *Applications of Systems Approaches at the Farm and Regional Levels. Systems Approaches for Sustainable Agricultural Development*. Kluwer, Dordrecht, pp. 1–18.

Jones, J.W., Hansen, J.W., Royce, F.S. and Messina, C.D. (2000) Potential benefits of climate forecasting to agriculture. *Agricultural Ecosystems and Environment* 82, 169–184.

Jordan, N., Mortensen, D.A., Prenzlow, D.M. and Cox, K.C. (1995) Simulation analysis of crop rotation effects on weed seedbanks. *American Journal of Botany* 82, 390–398.

Jordan, W.R., Dugas, W.A.J. and Shouse, P.J. (1983) Strategies for crop improvement for drought prone regions. *Agricultural Water Management* 7, 281–299.

Jovanovic, N.Z. and Annandale, J.G. (2000) Soil water balance: a computer tool for teaching future irrigation managers. *Journal of Natural Resources and Life Sciences Education* 29, 15–22.

Jovanovic, N.Z., Annandale, J.G. and Hammes, P.S. (2000) Teaching crop physiology with the soil water balance model. *Journal of Natural Resources and Life Sciences Education* 29, 23–30.

Kaku, M. (1998) *Visions – How Science will Revolutionize the 21st Century and Beyond.* Oxford University Press, Oxford.

Karim, Z., Ahmed, M., Hussain, S.G. and Rashid, K.B. (1991) *Impact of Climate Change on the Production of Modern Rice in Bangladesh.* Bangladesh Agricultural Research Council, Dhaka.

Keating, B.A., Wafula, B.M. and McCown, R.L. (1988) Simulation of plant density effects on maize yield as influenced by water and nitrogen limitations. In: Sinha, S.K., Shane, P.V., Bhargava, S.C. and Aggarwal, P.K. (eds) *Proceedings of the International Congress of Plant Physiology*, 15–20 February 1988, New Delhi. Society for Plant Physiology and Biochemistry, New Delhi.

Keating, B.A., Godwin, D.G. and Watiki, J.M. (1991) Optimising nitrogen inputs in response to climatic risk. In: Muchow, R.C. and Bellamy, J.A. (eds) *Climatic Risk in Crop Production – Models and Management for the Semi-arid Tropics and Subtropics.* CAB International, Wallingford, UK, pp. 329–357.

Keating, B.A., McCown, R.L. and Wafula, B.M. (1993) Adjustment of nitrogen inputs in response to a seasonal forecast in a region of high climatic risk. In: Penning de Vries, F.W.T., Teng, P. and Metselaar, K. (eds) *Systems Approaches for Agricultural Development.* Kluwer, Dordrecht, pp. 233–252.

Keatinge, J.D.H., Qi, A., Wheeler, T.R., Musitwa, F., Franks, N.A.P., Kuribuza, D., Ellis, R.H. and Sum- merfield, R.J. (1999) Potential annual-sown legumes for low-input systems in the East African Highlands of southwestern Uganda. *Mountain Research and Development* 19, 345–353.

Kebreab, E., France, J. and Ellis, R.H. (1998) *Evaluating the Potential of Biomathematical Modelling funded by DFID to Provide Developmental Impact in the Context of NRSP Goals and Purposes.* Department of Agriculture, University of Reading, Reading.

Keeling, C.D., Carter, A.F. and Mook, W.G. (1984) Seasonal, latitudinal, and secular variations in the abundance and isotopic ratios of atmospheric CO_2. *Journal of Geophysical Research* 89, 4615–4628.

Keen, P.G.W. and Scott-Morton, M.S. (1978) *Decision Support Systems: An Organisation's Perspective.* Addison-Wesley, Boston, Massachusetts.

Kelly, R.H., Parton, W.J., Crocker, G.J., Grace, P.R., Klír, J., Körschens, M., Poulton, P.R. and Richter, D.D. (1997) Simulating trends in soil organic carbon in long-term experiments using the CENTURY model. *Geoderma* 81, 75–90.

Kimball, B.A. (1983) Carbon dioxide and agricultural yield. *Agronomy Journal* 75, 779–788.

King, G.A. (1993) Conceptual approaches for incorporating climatic change into the development of forest management options for sequestering carbon. *Climate Research* 3, 61–78.

Kiniry, J.R., Williams, J.R., Gassman, P.W. and Debaeke, P. (1992) A general process-oriented model for two competing plant species. *Transactions of the ASAE* 35, 801–810.

Knight, J.D. (1997) The role of decision support systems in integrated crop protection. *Agriculture Ecosystems and Environment* 64, 157–163.

Knox, J., Matthews, R.B. and Wassmann, R. (2000) Using a crop/soil simulation model and GIS techniques to assess methane emissions from rice fields in Asia. III. Databases. *Nutrient Cycling in Agroecosystems* 58, 179–199.

Knox, J.W., Weatherhead, E.K. and Bradley, R.I. (1997) Mapping the total volumetric irrigation water requirements in England and Wales. *Agricultural Water Management* 33, 1–18.

Kramer, T., Ooijen, J.W. and Spitters, C.J.T. (1982) Selection for yield in small plots of wheat. *Euphytica* 31, 549–564.

Kropff, M.J. (1988) Modelling the effects of weeds on crop production. *Weed Research* 28, 465–471.

Kuyvenhoven, A., Ruben, R. and Kruseman, G. (1995) Options for sustainable agricultural systems and policy instruments to reach them. In: Bouma, J., Kuyvenhoven, A., Bouman, B.A.M., Luten, J.C. and Zandstra, H.G. (eds) *Eco-regional Approaches for Sustainable Land Use and Food Production. Systems Approaches for Sustainable Agricultural Development.* Kluwer, Dordrecht, pp. 187–212.

Lal, H., Hoogenboom, G., Calixte, J.P., Jones, J.W. and Beinroth, F.H. (1993) Using crop simulation models and GIS for regional productivity analysis. *Transactions of the ASAE* 36, 175–184.

Landivar, J.A., Baker, D.N. and Jenkins, J.N. (1983a) Application of GOSSYM to genetic feasibility studies. I. Analysis of fruit abscision and yield in okra-leaf cottons. *Crop Science* 23, 497–504.

Landivar, J.A., Baker, D.N. and Jenkins, J.N. (1983b) Application of GOSSYM to genetic feasibility stud-

ies. II. Analysis of increasing photosynthesis, specific leaf weight and longevity of leaves in cotton. *Crop Science* 23, 504–510.

Lansigan, F.P., Pandey, S. and Bouman, B.A.M. (1997) Combining crop modelling with economic risk-analysis for the evaluation of crop management strategies. *Field Crops Research* 51, 133–145.

Lau, L.J., Lin, W. and Yotopoulos, P.A. (1978) The linear logarithmic expenditure system: an application to consumption-leisure choice. *Econometrica* 46, 843–868.

Lawn, R.J. (1994) Exploiting physiology in crop improvement. *Plant Physiology Abstracts* 20, 467–476.

Lee, D.J., Tipton, T. and Leung, P. (1995) Modelling cropping decisions in a rural developing country: multiple objective programming approach. *Agricultural Systems* 49, 101–111.

Leemans, R. and Solomon, A.M. (1993) Modeling the potential change in yield and distribution of the earth's crops under a warmed climate. *Climate Research* 3, 79–96.

Li, C., Frolking, S. and Harriss, R.C. (1994) Modelling carbon biogeochemistry in agricultural soils. *Global Biogeochemical Cycles* 8, 237–254.

Li, C., Narayanan, V. and Harriss, R. (1996) Model estimates of nitrous oxide emissions from agricultural lands in the United States. *Global Biogeochemical Cycles* 10, 297–306.

Li, C., Zhuang, Y.H., Cao, M.Q., Crill, P.M., Dai, Z.H., Frolking, S., Moore, B., Salas, W., Song, W.Z. and Wang, X.K. (2001) Comparing a national inventory of N_2O emissions from arable lands in China developed with a process-based agro-ecosystem model to the IPCC methodology. *Nutrient Cycling in Agroecosystems* 60, 159–175.

Lindquist, J.L. and Kropff, M.J. (1997) Improving rice tolerance to banyardgrass through early crop vigour: simulations with INTERCOM. In: Kropff, M.J., Teng, P.S., Aggarwal, P.K., Bouma, J., Bouman, B.A.M., Jones, J.W. and van Laar, H.H. (eds) *Applications of Systems Approaches at the Field Level. Systems Approaches for Sustainable Agricultural Development*. Kluwer, Dordrecht, pp. 53–62.

Loomis, R.S. (1993) Optimization theory and crop improvement. In: Buxton, D.R., Shibles, R., Forsberg, R.A., Blad, B.L., Asay, K.H., Paulsen, G.M. and Wilson, R.F. (eds) *Proceedings of the First International Crop Science Congress*, Ames, Iowa, 14–22 July 1992. American Society of Agronomy, Madison, Wisconsin, pp. 583–588.

Loomis, R.S. and Williams, W.A. (1962) Maximum crop productivity: an estimate. *Crop Science* 3, 67–72.

Loveless, T. (1996) Why aren't computers used more in schools? *Educational Policy* 10, 1996.

Lu, R. and Siebenmorgen, T.J. (1994) Modeling rice field moisture content during the harvest season. Part II. Model implementation and validation. *Transactions of the ASAE* 37, 553–560.

Lu, R., Siebenmorgen, T.J., Dilday, R.H. and Costello, T.A. (1992) Modeling long-grain rice milling quality and yield during the harvest season. *Transactions of the ASAE* 35, 1905–1913.

Lundkvist, A. (1997) Weed management models. *Swedish Journal of Agricultural Research* 27, 155–166.

Luo, Y., TeBeest, D.O., Teng, P.S. and Fabellar, N.G. (1995) Simulation studies on risk analysis of rice leaf blast epidemics associated with global climate change in several Asian countries. *Journal of Biogeography* 22, 673–678.

Macadam, R., Britton, I., Russell, D., Potts, W., Baillie, B. and Shaw, A. (1990) The use of soft systems methodology to improve the adoption by Australian cotton growers of the Siratac computer based crop management system. *Agricultural Systems* 34, 1–14.

MacRobert, J.F. and Savage, M.J. (1998) The use of a crop simulation model for planning wheat irrigation in Zimbabwe. In: Tsuji, G.Y., Hoogenboom, G. and Thornton, P.K. (eds) *Understanding Options for Agricultural Production. Systems Approaches for Sustainable Agricultural Development*. Kluwer, Dordrecht, pp. 205–220.

MAFF (1983) Fertiliser recommendations for agricultural and horticultural crops. Research Bulletin 209. HMSO, London.

Magrin, G.O., Travasso, M.I., Diaz, R.A. and Rodriguez, R.O. (1997) Vulnerabilty of the agricultural systems in Argentina to climate change. *Climate Research* 9, 31–36.

Makin, I.W. and Cornish, G.A. (1996) Irrigation scheduling at system level: an analysis of practical applications of the INCA software. In: *Irrigation Scheduling from Theory to Practice*. FAO-ICID-CIID, Rome, pp. 279–287.

Malezieux, E., Zhang, J., Sinclair, E.R. and Bartholomew, D.P. (1994) Predicting pineapple harvest date in different environments, using a computer simulation model. *Agronomy Journal* 86, 609–617.

Martinez Garza, A. and Martinez Damian, M.A. (1996) Simulation model to forecast harvesting date for sugarcane based on temperature and precipitation data. *Agrociencia* 30, 487–494.

Matthews, R., Stephens, W., Hess, T.,

Mason, T. and Graves, A. (2002) Applications of crop/soil simulation models in tropical agricultural systems. *Advances in Agronomy* 76, 31–124.

Matthews, R.B. (2000) The role of quantitative methods in integrating biophysical and socio-economic aspects of soil fertility management. In: Ellis-Jones, J. (ed.) *Proceedings of a Workshop on Biophysical and Socio-economic Tools for Assessing Soil Fertility*, ARS-Lumle, Nepal, 17–18 October 2000. Silsoe Research Institute, Silsoe, UK.

Matthews, R.B. and Stephens, W. (1997) Evaluating irrigation strategies for tea using the CUPPA-Tea model. *Ngwazi Tea Research Unit Quarterly Report* 27, 3–11.

Matthews, R.B. and Stephens, W (1998a) CUPPA-Tea: a simulation model describing seasonal yield variation and potential production of tea. I. Shoot development and extension. *Experimental Agriculture* 34, 345-367.

Matthews, R.B. and Stephens, W. (1998b) The role of photoperiod in regulating seasonal yield variation in tea (*Camellia sinensis* L.). *Experimental Agriculture* 34, 323–340.

Matthews, R.B., Holden, S.T., Lungu, S. and Volk, J. (1992) The potential of alley cropping in improvement of cultivation systems in the high rainfall areas of Zambia. I. *Chitemene* and *fundikila. Agroforestry Systems* 17, 219–240.

Matthews, R.B., Kropff, M.J., Bachelet, D. and van Laar, H.H. (1995a) *Modeling the Impact of Climate Change on Rice Production in Asia.* IRRI/CAB International, Wallingford, UK.

Matthews, R.B., Kropff, M.J., Horie, T. and Bachelet, D. (1995b) Simulating the impact of climate change on rice production in Asia and evaluating options for adaptation. *Agricultural Systems* 54, 399–425.

Matthews, R.B., Wassmann, R. and Arah, J. (2000a) Using a crop/soil simulation model and GIS techniques to assess methane emissions from rice fields in Asia. I. Model development. *Nutrient Cycling in Agroecosystems* 58, 141–159.

Matthews, R.B., Wassmann, R., Buendia, L. and Knox, J. (2000b) Using a crop/soil simulation model and GIS techniques to assess methane emissions from rice fields in Asia. II. Model validation and sensitivity analysis. *Nutrient Cycling in Agroecosystems* 58, 161–177.

Matthews, R.B., Wassmann, R., Knox, J.W. and Buendia, L. (2000c) Using a crop/soil simulation model and GIS techniques to assess methane emissions from rice fields in Asia. IV. Upscaling of crop management scenarios to national levels. *Nutrient Cycling in Agroecosystems* 58, 201–217.

McAteer, E., Neil, D., Barr, N., Brown, M., Draper, S. and Henderson, F. (1996) Simulation software in a life sciences practical laboratory. *Computers and Education* 26, 101–112.

McCown, R.L. (2001) Learning to bridge the gap between science-based decision support and the practice of farming: evolution in paradigms of model-based research and intervention from design to dialogue. *Australian Journal of Agricultural Research* 52, 549–571.

McCown, R.L., Wafula, B.M., Mohammed, L., Ryan, J.G. and Hargreaves, J.N.G. (1991) Assessing the value of a seasonal rainfall predictor to agronomic decisions: the case of response farming in Kenya. In: Muchow, R.C. and Bellamy, J.A. (eds) *Climatic Risk in Crop Production – Models and Management for the Semi-arid Tropics and Sub-*

tropics. CAB International, Wallingford, UK, pp. 383–409.

McCown, R.L., Cox, P.G., Keating, B.A., Hammer, G.L., Carberry, P.S., Probert, M.E. and Freebairn, D.M. (1994) The development of strategies for improved agricultural systems and land-use management. In: Goldsworthy, P. and Penning de Vries, F.W.T. (eds) *Opportunities, Use, and Transfer of Systems Research Methods in Agriculture to Developing Countries. Systems Approaches for Sustainable Agricultural Development.* Kluwer, Dordrecht, pp. 81–96.

McDonagh, J. and Hillyer, A. (2000) Soil fertility, legumes and livelihoods in northern Namibia. In: Witcombe, J. and Harris, D. (eds) *DFID Plant Sciences Research Programme Annual Report 1999/2000.* Centre for Arid Zone Studies, University of Bangor, Bangor, UK.

McGlinchey, M.G., Inman-Bamber, N., Culverwell, T.L. and Els, M. (1995) An irrigation scheduling method based on a crop model and an automatic weather station. *Proceedings of the Annual Congress of the South African Sugar Technologists' Association* 69, 69–73.

McGregor, M.J., Rola-Rubzen, M.F. and Murray-Prior, R. (2001) Micro and macro-level approaches to modelling decision making. *Agricultural Systems* 69, 63–83.

McKinion, J.M., Baker, D.N., Whisler, F.D. and Lambert, J.R. (1989) Application of the GOSSYM/COMAX system to cotton management. *Agricultural Systems* 312, 55–65.

McLaren, J.S. and Craigon, J. (1981) A computer assisted learning system for use in the teaching of crop production. *Agricultural Progress* 56, 69–83.

McLaughlin, R.J. and Christie, B.R. (1980) Genetic variation for temperature response in alfalfa (*Medicago sativa* L.). *Canadian Journal of Plant Science* 60, 547–554.

Meinke, H. and Hammer, G.L. (1997) Forecasting regional crop production using SOI phases: an example for the Australian peanut industry. *Australian Journal of Agricultural Research* 48, 789–793.

Meinke, H., Stone, R.C. and Hammer, G.L. (1996) SOI phases and climatic risk to peanut production: a case study for northern Australia. *International Journal of Climatology* 16, 783–789.

Menz, K. and Grist, P. (1998) Bioeconomic modelling for analysing soil conservation policy issues. In: Penning de Vries, F.W.T., Agus, F. and Kerr, J. (eds) *Soil Erosion at Multiple Scales.* CAB International, Wallingford, UK, pp. 39–49.

Messina, C.D., Hansen, J.W. and Hall, A.J. (1999) Land allocation conditioned on El-Niño-Southern Oscillation phases in the Pampas of Argentina. *Agricultural Systems* 60, 197–212.

Metherell, A.K., Thorrold, B.S., Woodward, S.J.R., McCall, D.G., Marshall, P.R., Morton, J.D., Johns, K.L. and Baker, M. (1997) A decision support model for fertiliser recommendations for grazed pasture. *Proceedings of The New Zealand Grassland Association,* Mangere, Auckland, New Zealand, 29–31 October 1997, 59, 137–140.

Miller, A. (1993) The role of analytical science in natural resources decision-making. *Environmental Management* 17, 563–574.

Mohler, C.L. (1993) A model of the effects of tillage on emergence of weed seedlings. *Ecological Applications* 3, 53–73.

Monks, D.W. and Oliver, L.R. (1988) Interactions between soybean (*Glycine max*) cultivars and selected weeds. *Weed Science* 36, 770–774.

Monteith, J.L. (1981) Epilogue: themes and variations. *Plant and Soil* 58, 305–309.

Monteith, J.L. (1996) The quest for balance in crop modeling. *Agronomy Journal* 88, 695–697.

Morison, J.I.L. (1995) Software review: SB-ModelMaker for Windows, version 2.0b. *Agricultural and Forest Meteorology* 75, 265–267.

Muchow, R.C. and Carberry, P.S. (1993) Designing improved plant types for the semi-arid tropics: agronomists' viewpoints. In: Penning de Vries, F.W.T., Teng, P. and Metselaar, K. (eds) *Systems Approaches for Agricultural Development.* Kluwer, Dordrecht, pp. 37–61.

Muchow, R.C., Hammer, G.L. and Carberry, P.S. (1991) Optimising crop and cultivar selection in response to climatic risk. In: Muchow, R.C. and Bellamy, J.A. (eds) *Climatic Risk in Crop Production – Models and Management for the Semi-arid Tropics and Sub-tropics.* CAB International, Wallingford, UK, pp. 235–262.

Muchow, R.C., Hammer, G.L. and Vanderlip, R.L. (1994) Assessing climatic risk to sorghum production in water-limited sub tropical environments. II. Effect of planting date, soil water at planting, and cultivar phenology. *Field Crops Research* 36, 235–246.

Muchow, R.C., Cooper, M. and Hammer, G.L. (1996) Characterizing environmental challenges using models. In: Cooper, M. and Hammer, G.L. (eds) *Plant Adaptation and Crop Improvement.* CAB International, Wallingford, UK, pp. 349–364.

Murali, N.S., Secher, B.J.M., Rydahl, P. and Andreasen, F.M. (1999) Application of information technology in plant protection in Denmark: from vision to reality. *Computers and Electronics in Agriculture* 22, 109–115.

Mutsaers, H.J.W. and Wang, Z. (1999) Are simulation models ready for agricultural research in developing countries? *Agronomy Journal* 91, 1–4.

Narasimhan, T.N. (1995) Models and modeling of hydrogeologic processes. *Soil Science Society of America Journal* 59, 300–306.

NASS (2001) *Farm Computer Usage and Ownership.* National Agricultural Statistics Service, Washington, DC.

Nelson, R.A., Cramb, R.A. and Mamicpic, M.A. (1998a) Erosion/productivity modelling of maize farming in the Philippine uplands. III. Economic analysis of alternative farming methods. *Agricultural Systems* 58, 165–183.

Nelson, R.A., Dimes, J.P., Silburn, D.M., Paningbatan, E.P. and Cramb, R.A. (1998b) Erosion/productivity modelling of maize farming in the Philippine uplands. II. Simulation of alternative farming methods. *Agricultural Systems* 58, 147–163.

Newman, S., Lynch, T. and Plummer, A. (2000) Success and failure of decision support systems: Learning as we go. *Proceedings of the American Society of Animal Science* (E29), 1999. Available at: http://www.asas.org/jas/symposia/proceedings/0936.pdf (pp. 1–12).

Norton, R.D. (1995) Response to Section C. In: Bouma, J., Kuyvenhoven, A., Bouman, B.A.M., Luten, J.C. and Zandstra, H.G. (eds) *Eco-regional Approaches for Sustainable Land Use and Food Production. Systems Approaches for Sustainable Agricultural Development.* Kluwer, Dordrecht, pp. 237–247.

NSCGP (1992) *Ground for choices: Four Perspectives for Rural Areas in the*

European Community. Netherlands Scientific Council for Government Policy, 's-Gravenhage.

Okada, M. (1991) Variations of climate and rice production in northern Japan. In: Geng, S. and Cady, C.W. (eds) *Climate Variations and Change: Implications for Agriculture in the Pacific Rim.* University of California, Davis, California.

Omer, M.A., Saxton, K.E. and Bassett, D.L. (1988) Optimum sorghum planting dates in Western Sudan by simulated water budgets. *Agricultural Water Management* 13, 33–48.

Ortiz, R.A. (1998) Crop simulation models as an educational tool. In: Tsuji, G.Y., Hoogenboom, G. and Thornton, P.K. (eds) *Understanding Options for Agricultural Production. Systems Approaches for Sustainable Agricultural Development.* Kluwer, Dordrecht, pp. 383–388.

O'Toole, J.C. and Jones, C.A. (1987) Crop modelling: applications in directing and optimising rainfed rice research. *Weather and Rice: Proceedings of the International Workshop on the Impact of Weather Parameters on Growth and Yield of Rice.* International Rice Research Institute, Los Baños, The Philippines, pp. 255–269.

Palanisamy, S., Penning de Vries, F.W.T., Mohandass, S., Thiyagarajan, T.M. and Kareem, A.A. (1993) Simulation in pre-testing of rice genotypes in Tamil Nadu. In: Penning de Vries, F.W.T., Teng, P. and Metselaar, K. (eds) *Systems Approaches for Agricultural Development.* Kluwer, Dordrecht, pp. 63–75.

Panda, R.K., Stephens, W. and Matthews, R.B. (2002) Modelling the influence of irrigation on the potential yield of tea (*Camellia sinensis*) in NE India. *Experimental Agriculture* (in press).

Panigrahi, B. and Behara, B.P. (1998) Development of an irrigation calendar, based on crop–soil–climate modelling. *Indian Journal of Soil Conservation* 26, 273–279.

Pannell, D.J. (1996) Lessons from a decade of whole-farm modelling in Western Australia. *Review of Agricultural Economics* 18, 373–383.

Pannell, D.J. (1999) On the estimation of on-farm benefits of agricultural research. *Agricultural Systems* 61, 123–134.

Parton, W.J., Woomer, P.L. and Martin, A. (1994) Modelling soil organic matter dynamics and plant productivity in tropical ecosystems. In: Woomer, P.L. and Swift, M.J. (eds), *The Biological Management of Tropical Soil Fertility.* John Wiley & Sons, Chichester, UK, pp. 171–188.

Passioura, J.B. (1973) Sense and nonsense in crop simulation. *Journal of the Australian Institute of Agricultural Science* 39, 181–183.

Passioura, J.B. (1996) Simulation models: science, snake oil, education or engineering? *Agronomy Journal* 88, 690–694.

Peiris, T.S.G. and Thattil, R.O. (1998) The study of climate effects on the nut yield of coconut using parsimonious models. *Experimental Agriculture* 34, 189–206.

Penning de Vries, F.W.T. (1990) Can crop models contain economic factors? In: Rabbinge, R., Goudriaan, J., van Keulen, H., Penning de Vries, F.W.T. and van Laar, H.H. (eds) *Theoretical Production Ecology: Reflection and Prospects. Simulation Monographs.* Pudoc, Wageningen, pp. 89–103.

Penning de Vries, F.W.T. (1993) Rice production and climate change. In: Penning de Vries, F.W.T., Teng, P. and Metselaar, K. (eds) *Systems*

Approaches for Agricultural Development. Kluwer, Dordrecht, pp. 175–189.

Penning de Vries, F.W.T., Jansen, D.M., ten Berge, H.F.M. and Bakema, A. (1989) *Simulation of Ecophysiological Processes of Growth in Several Annual Crops. Simulation Monographs.* Pudoc/International Rice Research Institute, Wageningen.

Penning de Vries, F.W.T., van Keulen, H., van Diepen, C.A., Noy, I.G.A.M. and Goudriaan, J. (1990) Simulated yields of wheat and rice in current weather and in future weather when ambient CO_2 has doubled. Climate and Food Security, International Rice Research Institute, Los Baños, The Philippines, pp. 347–358.

Penning de Vries, F.W.T., van Keulen, H. and Rabbinge, R. (1995) Natural resources and limits of food production in 2040. In: Bouma, J., Kuyvenhoven, A., Bouman, B.A.M., Luten, J.C. and Zandstra, H.G. (eds) *Eco-regional Approaches for Sustainable Land Use and Food Production. Systems Approaches for Sustainable Agricultural Development.* Kluwer, Dordrecht, pp. 65–87.

Phadnawis, B.N. and Saini, A.D. (1992) Yield models in wheat based on sowing time and phenological developments. *Annals of Plant Physiology* 6, 52–59.

Philip, J.R. (1991) Soils, natural science and models. *Soil Science* 1511, 91–98.

Phillips, J.G., Cane, M.A. and Rosenzweig, C. (1998) ENSO, seasonal rainfall patterns and simulated maize yield variability in Zimbabwe. *Agricultural and Forestry Meteorology* 90, 39–50.

Pilkington, R. and Parker-Jones, C. (1996) Interacting with computer-based simulation: the role of dialogue.

Computers in Education 27, 1–14.

Pinnschmidt, H.O., Teng, P.S., Yuen, J.E. and Djurle, A. (1990) Coupling pest effects to the IBSNAT CERES crop model for rice. *Phytopathology* 80, 997.

Pinnschmidt, H.O., Chamarerk, V., Cabulisan, N., Dela Peña, F., Long, N.D., Savary, S., Klein-Gebbinck, H.W. and Teng, P.S. (1997) Yield gap analysis of rainfed lowland systems to guide rice crop and pest management. In: Kropff, M.J., Teng, P.S., Aggarwal, P.K., Bouma, J., Bouman, B.A.M., Jones, J.W. and van Laar, H.H. (eds) *Applications of Systems Approaches at the Field Level. Systems Approaches for Sustainable Agricultural Development.* Kluwer, Dordrecht, pp. 321–338.

Pinstrup-Andersen, P., Pandya-Lorch, R. and Rosegrant, M.W. (1999) *World Food Prospects: Critical Issues for the Early Twenty-First Century.* Food Policy Report. International Food Policy Research Institute, Washington, DC.

Piper, E.L., Boote, K.J. and Jones, J.W. (1993) Environmental effects of oil, protein composition of soybean seed. *Agronomy Abstracts* 1993, 152–153.

Plauborg, F. and Heidmann, T. (1996) MARKVAND: an irrigation scheduling system for use under limited irrigation capacity in a temperate humid climate. In: *Irrigation Scheduling from Theory to Practice (Water Reports).* FAO-ICID-CIID, Rome, pp. 177–184.

Prins, A., Bornman, J.J. and Meyer, J.H. (1997) Economic fertiliser recommendations for sugarcane in Kwazulu-Natal, incorporating risk quantification using the KYNOCANE computer program. *Proceedings of the Annual Congress South African Sugar Technologists' Association* No. 71, pp. 38–41.

Probert, M.E., Keating, B.A., Thompson, J.P. and Parton, W.J. (1995) Modelling water, nitrogen and crop yield for a long-term fallow management experiment. *Australian Journal of Experimental Agriculture* 35, 941–950.

Probert, M.E., Dimes, J.P., Keating, B.A., Dalal, R.C. and Strong, W.M. (1998) APSIM's water and nitrogen modules and simulation of the dynamics of water and nitrogen in fallow systems. *Agricultural Systems* 56, 1–28.

Rabbinge, R. and Rijsdijk, F.H. (1983) EPIPRE: a disease and pest management system for winter wheat, taking account of micrometeorological factors. *EPPO Bulletin* 13, 297–305.

Rabbinge, R., Leffelaar, P.A. and van Latesteijn, H.C. (1994) The role of systems analysis as an instrument in policy making and resource management. In: Goldsworthy, P. and Penning de Vries, F.W.T. (eds) *Opportunities, Use, and Transfer of Systems Research Methods in Agriculture to Developing Countries. Systems Approaches for Sustainable Agricultural Development.* Kluwer, Dordrecht, pp. 67–79.

Rasmusson, D.C. (1991) A plant breeder's experience with ideotype breeding. *Field Crops Research* 26, 191–200.

Rimmington, G.M. (1984) A model of the effect of interspecies competition for light on dry-matter production. *Australian Journal of Plant Physiology* 11, 277–286.

Ritchie, J.T., Singh, U., Godwin, D.C. and Bowen, W.T. (1998) Cereal growth, development and yield. In: Tsuji, G.Y., Hoogenboom, G. and Thornton, P.K. (eds) *Understanding Options for Agricultural Production. Systems Approaches for Sustainable Agricultural Develop-*ment. Kluwer, Dordrecht, pp. 79–98.

Robertson, D., Johnston, W. and Nip, W. (1995) Virtual frog dissection – interactive 3D graphics via the web. *Computer Networks and ISDN Systems* 28, 155–160.

Robertson, M.J., Carberry, P.S. and Lucy, M. (2000) Evaluation of a new cropping option using a participatory approach with on-farm monitoring and simulation: a case study of spring-sown mungbeans. *Australian Journal of Agricultural Research* 51, 1–12.

Roetter, R.P., van Keulen, H., Laborte, A.G., Hoanh, C.T. and van Laar, H.H. (eds) (2000a) *Systems Research for Optimising Future Land Use in South and Southeast Asia.* SysNet Research Paper Series No. 2, International Rice Research Institute, Makati, The Philippines.

Roetter, R.P., van Keulen, H. and van Laar, H.H. (eds) (2000b) *Synthesis of Methodology Development and Case Studies.* SysNet Research Paper Series No. 3, International Rice Research Institute, Makati, The Philippines.

Rosenzweig, C., Parry, M.L., Fischer, G. and Frohberg, K. (1993) *Climate Change and World Food Supply, 3.* Environmental Change Unit, University of Oxford, Oxford.

Saseendran, S.A., Hubbard, K.G., Singh, K.K., Mendiratta, N., Rathore, L.S. and Singh, S.V. (1998) Optimum transplanting dates for rice in Kerala, India, determined using both CERES v3.0 and ClimProb. *Agronomy Journal* 90, 185–190.

Satake, T. and Yoshida, S. (1978) High temperature-induced sterility in *indica* rice at flowering. *Japanese Journal of Crop Science* 47, 6–17.

Savin, R., Satorre, E.H., Hall, A.J. and Slafer, G.A. (1995) Assessing strategies for wheat cropping in the monsoonal climate of the Pampas using the CERES-Wheat simulation model. *Field Crops Research* 42, 81–91.

Schaber, J. (1997) FARMSIM: a dynamic model for the simulation of yields, nutrient cycling and resource flows on Philippine small-scale farming systems. MSc Thesis, Universität Osnabrück, Osnabrück.

Schipper, R., Jansen, H.G.P., Bouman, B.A.M., Hengsdijk, H., Nieuenhuyse, A. and Sáenz, F. (2001) Integrated bioeconomic land-use models: an analysis of policy issues in the Atlantic Zone of Costa Rica. In: Lee, D.R. and Barrett, C.B. (eds) *Tradeoffs or Synergies? Agricultural Intensification, Economic Development, and the Environment.* CAB International, Wallingford, UK, pp. 267–284.

Schipper, R.A., Jansen, H.G.P., Stoorvogel, J.J. and Jansen, D.M. (1995) Evaluating policies for sustainable land use: a subregional model with farm types in Costa Rica. In: Bouma, J., Kuyvenhoven, A., Bouman, B.A.M., Luten, J.C. and Zandstra, H.G. (eds) *Eco-regional Approaches for Sustainable Land Use and Food Production. Systems Approaches for Sustainable Agricultural Development.* Kluwer, Dordrecht, pp. 377–395.

Schon, D.A. (1983) *The Reflective Practitioner: How Professionals Think in Action.* Basic Books, New York.

Schouwenaars, J.M. and Pelgrum, G.H. (1990) A model approach to analyse sowing strategies for maize in southern Mozambique. *Netherlands Journal of Agricultural Science* 38, 9–20.

Scott, R. and Robinson, B. (1996) Managing technological change in education – what lessons can we all learn? *Computers and Education* 26, 131–134.

Seligman, N.G. (1990) The crop model record: promise or poor show? In: Rabbinge, R., Goudriaan, J., van Keulen, H., Penning de Vries, F.W.T. and van Laar, H.H. (eds) *Theoretical Production Ecology: Reflection and Prospects.* Simulation Monographs, Pudoc, Wageningen, pp. 249–263.

Seligman, N.G. and van Keulen, H. (1981) PAPRAN: a simulation model of annual pasture production limited by rainfall and nitrogen. In: Frissel, M.J. and van Veen, J.A. (eds) *Simulation of Nitrogen Behaviour in Soil–plant Systems.* PUDOC, Wageningen, pp. 192–221.

Selvarajan, S., Aggarwal, P.K., Pandey, S., Lansigan, F.P. and Bandyopadhyay, S.K. (1997) Systems approach for analysing tradeoffs between income, risk, and water use in rice–wheat production in northern India. *Field Crops Research* 51, 147–161.

Shane, W., Teng, P.S., Laney, A. and Cattanach, A. (1985) Management of *Cercospora* leafspot in sugarbeets: decision aids. *North Dakota Farm Research* 43, 3–5.

Sheehy, J.E., Bergersen, F.J., Minchin, F.R. and Witty, J. (1987) A simulation study of gaseous diffusion resistance, nodule pressure gradients and biological nitrogen fixation in soyabean nodules. *Annals of Botany* 60, 345–351.

Shepherd, K.D. and Soule, M.J. (1998) Soil fertility management in west Kenya: a dynamic simulation of productivity, profitability and sustainability at different resource endowment levels. *Agricultural, Ecosystems and Environment* 71, 131–145.

Shorter, R., Lawn, R.J. and Hammer, G.L. (1991) Improving genotypic adaptation in crops – a role for breeders, physiologists and modellers. *Experimental Agriculture* 27, 155–175.

Sichman, J.S., Conte, R. and Gilbert, N. (1998) Multi-agent systems and agent-based modelling. Springer, Berlin.

Simmonds, L., Schofield, J. and Mullins, C. (1995) SPACTeach – a computer assisted learning module exploring water movement in the soil–plant–atmosphere continuum. MERTaL Courseware, University of Aberdeen, Aberdeen.

Simon, H.A. (1979) Rational decision making in business organisations. *American Economic Review* 69, 493–513.

Sinclair, T.R. (1986) Water and nitrogen limitations in soybean grain production I. Model development. *Field Crops Research* 15, 125–141.

Sinclair, T.R. and Seligman, N.A.G. (1996) Crop modelling: from infancy to maturity. *Agronomy Journal* 88, 698–703.

Singels, A. (1992) Evaluating wheat planting strategies using a growth model. *Agricultural Systems* 38, 175–184.

Singels, A. (1993) Determining optimal flowering and sowing dates for wheat in the central irrigation areas of the RSA using a growth model. *South African Journal of Plant and Soil* 10, 77–84.

Singels, A. and Bezuidenhout, C.N. (1998) ENSO, the South African climate and sugarcane production. *Proceedings of the South African Sugar Technologists' Association* 72, 10–11.

Singels, A. and Bezuidenhout, C.N. (1999) The relationship between ENSO and rainfall and yield in the South African sugar industry. *South African Journal of Plant and Soil* 16, 96–101.

Singels, A. and Potgieter, A.B. (1997) A technique to evaluate ENSO-based maize production strategies. *South African Journal of Plant and Soil* 14, 93–97.

Singels, A., Kennedy, A.J. and Bezouidenhout, C.N. (1998) IRRICANE: a simple computerised irrigation sceduling method for sugarcane. *Proceedings of the South African Sugar Technologist's Association* 72, 117–122.

Singels, A., Kennedy, A.J. and Bezouidenhout, C.N. (2000) Weather based decision support through the internet for agronomic management of sugar cane. *Proceedings of the South African Sugar Technology Association* 73, 30–32.

Singh, P., Boote, K.J., Rao, A.Y. and Iruthayaraj, M.R. (1994) Evaluation of the groundnut model PNUTGRO for crop response to water availability, sowing dates, and seasons. *Field Crops Research* 39, 147–162.

Singh, P., Alagarswamy, G., Hoogenboom, G., Pathak, P., Wani, S.P. and Virmani, S.M. (1999) Soybean–chickpea rotation on Vertic Inceptisols. II. Long-term simulation of water balance and crop yields. *Field Crops Research* 63, 225–236.

Singh, U. and Thornton, P.K. (1992) Using crop models for sustainability and environmental quality assessment. *Outlook on Agriculture* 21, 209–218.

Singh, U., Thornton, P.K., Saka, A. and Dent, J.B. (1993) Maize modelling in Malawi: a tool for soil fertility research and development. In: Penning de Vries, F.W.T., Teng, P. and Metselaar, K. (eds), *Systems Approaches for Agricultural Development*. Kluwer, Dordrecht, pp. 253–273.

Smeets, E., Vandenriessche, H., Hendrickx, G., de Wijngaert, K. and Geypens, M. (1992) Photosanitary balance of winter wheat in 1992 by the EPIPRE advice system. *Parasitica* 48, 139–148.

Smith, J.U., Dailey, A.G., Glendining, M.J., Bradbury, N.J., Addiscott, T.M., Smith, P., Bide, A., Boothroyd, D., Brown, E., Cartwright, R., Chorley, R., Cook, S., Cousins, S., Draper, S., Dunn, M., Fisher, A., Griffith, P., Hayes, C., Lock, A., Lord, S., Mackay, J., Malone, C., Mitchell, D., Nettleton, D., Nicholls, D. and Overman, H. (1997a) Constructing a nitrogen fertilizer recommendation system using a dynamic model: what do farmers want? *Soil Use and Management* 13, 225–228.

Smith, M.C., Holt, J. and Webb, M. (1993) Population model of the parasitic weed *Striga hermonthica* (*Scrophulariaceae*) to investigate the potential of *Smicronyx umbrinus* (Coleoptera: *Curculionidae*) for biological control in Mali. *Crop Protection* 12, 470–476.

Smith, P., Smith, J.U., Powlson, D.S., McGill, W.B., Arah, J.R.M., Chertov, O.G., Coleman, K., Franko, U., Frolking, S., Jenkinson, D.S., Jensen, L.S., Kelly, R.H., Klein-Gunnewiek, H., Komarov, A.S., Li, C., Molina, J.A.E., Mueller, T., Parton, W.J., Thornley, J.H.M. and Whitmore, A.P. (1997b) A comparison of the performance of nine organic matter models using datasets from seven long-term experiments. *Geoderma* 81, 153–225.

Solomon, A.M. and Leemans, R. (1990) Climatic change and landscape-ecological response: issues and analysis. In: Boer, M.M. and de Groot, R.S. (eds) *Landscape-Ecological Impact of Climate Change*. IOS Press, Amsterdam, pp. 293–316.

Spedding, C.R.W. (1990) Agricultural production systems. In: Rabbinge, R., Goudriaan, J., van Keulen, H., Penning de Vries, F.W.T. and van Laar, H.H. (eds) *Theoretical Pro-*

duction Ecology: Reflection and Prospects. Simulation Monographs, Pudoc, Wageningen, pp. 239–248.

Spitters, C.J.T. and Aerts, R. (1983) Simulation of competition for light and water in crop-weed associations. *Aspects of Applied Biology* 4, 467–484.

Squire, G.R. (1985) Ten years of tea physiology. *Tea* 8, 42–48.

Stange, F., Butterbach-Bahl, K., Papen, H., Zechmeister-Boltenstern, S., Li, C. and Aber, J. (2000) A process-oriented model of N_2O and NO emission from forest soils. 2. Sensitivity analysis and validation. *Journal of Geophysical Research* 105, 4385–4398.

Stapper, M. and Harris, H.C. (1989) Assessing the productivity of wheat genotypes in a mediterranean climate using a crop simulation model. *Field Crops Research* 20, 129–152.

Stephens, W. and Hess, T. (1996) Report on the PARCH evaluation visit to Kenya, Malawi, Zimbabwe and Botswana. Cranfield University, Silsoe, UK.

Stephens, W. and Hess, T.M. (1999) Modelling the benefits of soil water conservation using the PARCH model – a case study from a semi-arid region of Kenya. *Journal of Arid Environments* 41, 335–344.

Stephens, W., Hess, T.M., Crout, N.M.J., Young, S.D. and Bradley, R.G. (1996) *PARCH – Tutorial Guide*. Natural Resources Institute, Chatham, UK.

Stewart, J.I. (1991) Principles and performance of response farming. In: Muchow, R.C. and Bellamy, J.A. (eds) *Climatic Risk in Crop Production – Models and Management for the Semi-arid Tropics and Sub-tropics*. CAB International, Wallingford, UK, pp. 361–382.

Stickler, F.C. and Wearden, S. (1965)

Yield and yield components of grain sorghum as affected by row width and stand density. *Agronomy Journal* 57, 564–567.

Stockle, C.O., Martin, S.A. and Campbell, G.S. (1994) CropSyst, a cropping systems simulation model: water/nitrogen budgets and crop yield. *Agricultural Systems* 46, 335–359.

Stoorvogel, J.J. (1995) Integration of computer-based models and tools to evaluate alternative land-use scenarios as part of an agricultural systems analysis. *Agricultural Systems* 49, 353–367.

Stroosnijder, L. and Kiepe, P. (1997) Systems approach in the design of soil and water conservation measures. In: Kropff, M.J., Teng, P.S., Aggarwal, P.K., Bouma, J., Bouman, B.A.M., Jones, J.W. and van Laar, H.H. (eds) *Applications of Systems Approaches at the Field Level. Systems Approaches for Sustainable Agricultural Development.* Kluwer, Dordrecht, pp. 399–411.

Stuth, J.W. and Lyons, B.G. (eds) (1993) *Decision Support Systems for the Management of Grazing lands. Man and the Biosphere,* II. UNESCO, Paris.

Surin, A., Rojanahusdin, P., Munkong, W., Dhitikiattinpong, S.R. and Disthaporn, S. (1991) Using empirical blast models to establish disease management recommendations in Thailand, Rice Blast Modelling and Forecasting. IRRI, Los Baños, The Phillippines, pp. 69–74.

ten Berge, H.F.M. (1993) Building capacity for systems research at national agricultural research centres: SARP's experience. In: Penning de Vries, F.W.T., Teng, P. and Metselaar, K. (eds) *Systems Approaches for Agricultural Development.* Kluwer, Dordrecht, pp. 515–538.

ten Berge, H.F.M., Kropff, M.J. and Wopereis, M.C.S. (1994) *The SARP project Phase III (1992–1995): Overview, Goals, Plans.* Wageningen Agricultural University, Wageningen.

ten Berge, H.F.M., Shi, Q., Zheng, Z., Rao, K.S., Riethoven, J.J.M. and Zhong, X. (1997a) Numerical optimisation of nitrogen application to rice. II. Field evaluations. *Field Crops Research* 51, 43–54.

ten Berge, H.F.M., Thiyagarajan, T.M., Shi, Q., Wopereis, M.C.S., Drenth, H. and Jansen, M.J.W. (1997b) Numerical optimisation of nitrogen application to rice. I. Description of MANAGE-N. *Field Crops Research* 51, 29–42.

Teng, P.S. (1987) The systems approach to pest management. In: Teng, P.S. (ed.) *Crop Loss Assessment and Pest Management.* American Phytopathological Society Press, St Paul, Minnesota, pp. 160–167.

Teng, P.S. and Savary, S. (1992) Implementing the systems approach in pest management. *Agricultural Systems* 40, 237–264.

Teng, P.S., Blackie, M.J. and Close, R.C. (1978) Simulation modeling of plant diseases to rationalize fungicide use. *Outlook on Agriculture* 9, 273–277.

Teng, P.S., Batchelor, W.D., Pinnschmidt, H.O. and Wilkerson, G.G. (1998) Simulation of pest effects on crops using coupled pest-crop models: the potential for decision support. In: Tsuji, G.Y., Hoogenboom, G. and Thornton, P.K. (eds) *Understanding Options for Agricultural Production. Systems Approaches for Sustainable Agricultural Development.* Kluwer, Dordrecht, pp. 221–266.

Thiyagarajan, T.M., Stalin, P., Dobermann, A., Cassman, K.G. and ten Berge, H.F.M. (1997) Soil N supply and plant N uptake by irrigated rice in Tamil Nadu. *Field Crops Research* 51, 55–64.

Thomas, R. and Neilson, I. (1995) Harnessing simulations in the service of education: the Interact simulation environment. *Computers in Education* 25(1/2), 21–29.

Thornton, P.K. and Jones, P.G. (1997) Towards a conceptual dynamic land-use model. In: Teng, P.S., Kropff, M.J., ten Berge, H.F.M., Dent, J.B., Lansigan, F.P. and van Laar, H.H. (eds) *Applications of Systems Approaches at the Farm and Regional Levels. Systems Approaches for Sustainable Agricultural Development.* Kluwer, Dordrecht, pp. 341–356.

Thornton, P.K. and Wilkens, P.W. (1998) Risk assessment and food security. In: Tsuji, G.Y., Hoogenboom, G. and Thornton, P.K. (eds) *Understanding Options for Agricultural Production. Systems Approaches for Sustainable Agricultural Development.* Kluwer, Dordrecht, pp. 329–345.

Thornton, P.K., Fawcett, R.H., Dent, J.B. and Perkins, T.J. (1990) Spatial weed distribution and economic thresholds for weed control. *Crop Protection* 9, 337–342.

Thornton, P.K., Hansen, J.W., Knapp, E.B. and Jones, J.W. (1995a) Designing optimal crop management strategies. In: Bouma, J., Kuyvenhoven, A., Bouman, B.A.M., Luten, J.C. and Zandstra, H.G. (eds) *Eco-regional Approaches for Sustainable Land Use and Food Production. Systems Approaches for Sustainable Agricultural Development.* Kluwer, Dordrecht, pp. 333–351.

Thornton, P.K., Saka, A.R., Singh, U., Kumwenda, J.D.T., Brink, J.E. and Dent, J.B. (1995b) Application of a maize crop simulation model in the central region of Malawi. *Experimental Agriculture* 31, 213–226.

Thornton, P.K., Bowen, W.T., Ravelo, A.C., Wilkens, P.W., Farmer, G., Brock, J. and Brink, J.E. (1997) Estimating millet production for famine early warning: an application of crop simulation modelling using satellite and ground-based data in Burkina Faso. *Agricultural and Forest Meteorology* 83, 95–112.

Timsina, J., Singh, U. and Singh, Y. (1997) Addressing sustainability of rice-wheat systems: analysis of long-term experimentation and simulation. In: Kropff, M.J., Teng, P.S., Aggarwal, P.K., Bouma, J., Bouman, B.A.M., Jones, J.W. and van Laar, H.H. (eds) *Applications of Systems Approaches at the Field Level. Systems Approaches for Sustainable Agricultural Development.* Kluwer, Dordrecht, pp. 383–397.

Timsina, J., Godwin, D. and Connor, D.J. (2001) Water management for mungbean within rice-wheat cropping systems of Bangladesh: a simulation study. In: *Proceedings – Third International Symposium on Systems Approaches for Agricultural Development,* Lima, 8–10 November 1999. International Potato Centre (CIP), Lima.

Tollefson, L. (1996) Requirements for improved interactive communication between researchers, managers, extensionists and farmers. In: *Irrigation Scheduling from Theory to Practice (Water Reports).* FAO-ICID-CIID, Rome, pp. 217–226.

Tsuji, G. and Balas, S. (eds) (1993) *The IBSNAT Decade: Ten Years of Endeavour at the Frontier of Science and Technology.* Department of Agronomy and Soil Science, University of Hawaii, College of Tropical Agriculture and Human Resources, Honolulu, Hawaii.

Tsuji, G.Y., Hoogenboom, G. and Thornton, P.K. (1998) *Understanding Options for Agricultural Production*. Kluwer, Dordrecht.

Uehara, G. and Tsuji, G.Y. (1993) The IBSNAT Project. In: Penning de Vries, F.W.T., Teng, P. and Metselaar, K. (eds) *Systems Approaches for Agricultural Development*. Kluwer, Dordrecht.

UNEP (1999) *Global Environmental Outlook 2000 – UNEP's Millennium Report on the Environment*. Earthscan Publications, London.

UNEP/SCOPE (1999) *Emerging Environmental Issues for the 21st Century: a study for GEO 2000*. UNEP/DE1A&EW/TR. 99–5, Scientific Committee on Problems of the Environment, Paris.

van der Meer, F.-B.W., Twomlow, S.J., Bruneau, P.M.C. and Reid, I. (1999) Weed management in semi-arid agriculture: application of a soil moisture competition model. *Proceedings of the Brighton Weeds Conference*, Brighton 15–19 November 1999. BCPC, Farnham, UK.

van Diepen, C.A., Rappoldt, C., Wolf, J. and van Keulen, H. (1988) *Crop Growth Simulation Model WOFOST*. Centre for World Food Studies, Wageningen.

van Hofwegen, P. (1996) Social, cultural, institutional and policy constraints. In: *Irrigation Scheduling from Theory to Practice (Water Reports)*. FAO-ICID-CIID, Rome, pp. 323–333.

van Keulen, H. (1975) *Simulation of Water use and Herbage Growth in Arid Regions*. Simulation Monographs, Pudoc, Wageningen.

van Keulen, H. (1993) Options for agricultural development: a new quantitative approach. In: Penning de Vries, F.W.T., Teng, P. and Metselaar, K. (eds) *Systems Approaches for Agricultural Development*. Kluwer, Dordrecht, pp. 355–365.

van Keulen, H. (1995) Sustainability and long-term dynamics of soil organic matter and nutrients under alternative management strategies. In: Bouma, J., Kuyvenhoven, A., Bouman, B.A.M., Luten, J.C. and Zandstra, H.G. (eds) *Eco-regional Approaches for Sustainable Land Use and Food Production. Systems Approaches for Sustainable Agricultural Development*. Kluwer, Dordrecht, pp. 353–375.

van Keulen, H. and Stol, W. (1995) Agro-ecological zonation for potato production. In: Haverkort, A.J. and MacKerron, D.K.L. (eds) *Potato Ecology and Modelling of Crops Under Conditions Limiting Growth*. Kluwer, Dordrecht, pp. 357–371.

van Keulen, H. and Veeneklaas, F.R. (1993) Options for agricultural development: a case study for Mali's fifth Region. In: Penning de Vries, F.W.T., Teng, P. and Metselaar, K. (eds) *Systems Approaches for Agricultural Development*. Kluwer, Dordrecht, pp. 367–380.

van Latesteijn, H.C. (1993) A methodological framework to explore long-term options for land-use. In: Penning de Vries, F.W.T., Teng, P. and Metselaar, K. (eds) *Systems Approaches for Agricultural Development*. Kluwer, Dordrecht, pp. 445–455.

van Lenteren, J.C. and van Roermund, H.J.W. (1997) Better biological control by a combination of experimentation and modelling. In: Kropff, M.J., Teng, P.S., Aggarwal, P.K., Bouma, J., Bouman, B.A.M., Jones, J.W. and van Laar, H.H. (eds) *Applications of Systems Approaches at the Field Level. Systems Approaches for Sustainable Agricultural Development*. Kluwer, Dordrecht, pp. 349–363.

van Ranst, E. and Vanmechelen, L. (1995) Application of computer-based technology in assessing production potentials for annual crops in developing countries. *Bulletin des Séances. Académie royale des Sciences d'outre-mer, Brussels* 41, 613–636.

VanDevender, K.W., Costello, T.A. and Smith, R.J., Jr (1997) Model of rice (*Oryza sativa*) yield reduction as a function of weed interference. *Weed Science* 45, 218–224.

Verburg, K., Keating, B.A., Bristow, K.L., Huth, N.I., Ross, P.J. and Catchpoole, V.R. (1996) Evaluation of nitrogen fertiliser management strategies in sugarcane using APSIM-SWIM. In: Wilson, J.R., Hogarth, D.M., Campbell, J.A. and Garside, A.L. (eds) *Sugarcane: Research Towards Efficient and Sustainable Production.* CSIRO Division of Tropical Crops and Pastures, Brisbane, pp. 200–202.

Wade, L.J. (1991) Optimising plant stand in response to climatic risk. In: Muchow, R.C. and Bellamy, J.A. (eds) *Climatic Risk in Crop Production – Models and Management for the Semi-arid Tropics and Sub-tropics.* CAB International, Wallingford, UK, pp. 263–282.

Wafula, B.M. (1995) Applications of crop simulation in agricultural extension and research in Kenya. *Agricultural Systems* 49, 399–412.

Walker, D.H. and Lowes, D. (1997) Natural resource management: opportunities and challenges in the application of decision support systems. *AI Applications* 11, 41–51.

Wang, F., Shili, W., Yuxiang, L. and Meina, Z. (1991) A preliminary modelling of the effects of climate changes on food production in China. In: Geng, S. and Cady, C.W. (eds) *Climate Variations and Change: Implications for Agriculture in the Pacific Rim.* University of California, Davis, California, pp. 115–126.

Watanabe, T., Fabellar, L.T., Almazan, L.P., Rubia, E.G., Heong, K.L. and Sogawa, K. (1997) Quantitative evaluation of growth and yield of rice plants infested with rice planthoppers. In: Kropff, M.J., Teng, P.S., Aggarwal, P.K., Bouma, J., Bouman, B.A.M., Jones, J.W. and van Laar, H.H. (eds) *Applications of Systems Approaches at the Field Level. Systems Approaches for Sustainable Agricultural Development.* Kluwer, Dordrecht, pp. 365–382.

Way, M.J. and van Emden, H.F. (2000) Integrated pest management in practice – pathways towards successful application. *Crop Protection* 19, 81–103.

Welch, S.M., Jones, J.W., Reeder, G., Brennan, M.W. and Jackson, B.M. (1999) PCYIELD: Model-based support for soybean production. In: Donatelli, M., Stockle, C., Villalobos, F. and Mir, J.M.V. (eds) *Proceedings of the International Symposium on Modelling Cropping Systems, 21–23 June 1999, Lleida, Spain.* University of Lleida, Lleida, Spain, pp. 273–274.

Wheeler, T.R., Qui, A., Keatinge, J.D.H., Ellis, R.H. and Summerfield, R.J. (1999) Selecting legume cover crops for hillside environments in Bolivia. *Mountain Research and Development* 19, 318–324.

Whisler, F.D., Acock, B., Baker, D.N., Fye, R.E., Hodges, H.F., Lambert, J.R., Lemmon, H.E., McKinion, J.M. and Reddy, V.R. (1986) Crop simulation models in agronomic systems. *Advances in Agronomy* 40, 141–208.

White, J.W. (1998) Modelling and crop improvement. In: Tsuji, G.Y., Hoogenboom, G. and Thornton, P.K. (eds) *Understanding Options for Agricultural Production. Systems Approaches for Sustainable Agricultural Development.* Kluwer, Dordrecht, pp. 179–188.

Whittaker, A.D. (1993) Decision support systems and expert systems for range science. In: Stuth, J.W. and Lyons, B.G. (eds) *Decision Support Systems for the Management of Grazing Lands. Man and the Biosphere.* UNESCO, Paris, pp. 69–82.

Wiles, L.J. and Wilkerson, G.G. (1991) Modelling competition for light between soybean and broadleaf weeds. *Agricultural Systems* 35, 37–51.

Wilkerson, G.G., Jones, J.W., Boote, K.J., Ingram, K.I. and Mishoe, J.W. (1983) Modelling soybean growth for crop management. *Transactions of the ASAE* 26, 63–73.

Williams, J.R., Jones, C.A. and Dyke, P.T. (1984) A modelling approach to determining the relationship between erosion and soil productivity. *Transactions of the American Society of Engineers* 27, 129–144.

Williams, J.R., DeLano, D.R., Heiniger, R.W., Vanderlip, R.L. and Llewelyn, R.V. (1999) Replanting strategies for grain sorghum under risk. *Agricultural Systems* 60, 137–155.

Williams, T.C. and Zahed, H. (1996) Computer-based training versus traditional lecture: effect on learning and retention. *Journal of Business and Psychology* 11, 297–310.

Wolf, J., Berkout, J.A.A., van Diepen, C.A. and van Immerzeel, C.H. (1989) A study on the limitations to maize production in Zambia using simulation models and a geographic information system. In: Bouma, J. and Bregt, A.K. (eds) *Land Qualities in Space and Time: Proceedings of a Symposium Organised by the ISSS*, Wageningen, 22–26 August 1988. Pudoc, Wageningen.

Yin, X., Kropff, M.J., Aggarwal, P.K.,

Peng, S. and Horie, T. (1997) Optimal preflowering phenology of irrigated rice for high yield potential in three Asian environments: a simulation study. *Field Crops Research* 51, 19–27.

Yin, X., Kropff, M.J. and Stam, P. (1999a) The role of ecophysiological models in QTL analysis: the example of specific leaf area in barley. *Heredity* 82, 415–421.

Yin, X., Stam, P., Dourleijn, C.J. and Kropff, M.J. (1999b) AFLP mapping of quantitative trait loci for yield-determining physiological characters in spring barley. *Theoretical and Applied Genetics* 99, 244–253.

Yin, X., Kropff, M.J., Goudriaan, J. and Stam, O. (2000) A model analysis of yield differences among recombinant inbred lines in barley. *Agronomy Journal* 92, 114–120.

Yoshino, M.M., Horie, T., Seino, H., Tsujii, H., Uchijima, T. and Uchijima, Z. (1988) The effect of climatic variations on agriculture in Japan. In: Parry, M.L., Carter, T.R. and Konijn, N.T. (eds) *The Impact of Climate Variations on Agriculture. Volume 1: Assessments in Cool Temperate and Cold Regions.* Kluwer, Dordrecht, pp. 853–868.

Yotopoulos, P.A., Lau, L.J. and Lin, W. (1976) Microeconomic output supply and factor demand functions in the agriculture of the Province of Taiwan. *American Journal of Agricultural Economics* 58, 333–340.

Young, C. and Heath, S.B. (1991) Computer assisted learning in agriculture and forestry education – the way forward. *Agricultural Progress* 66, 56–63.

Young, M.D.B. and Gowing, J.W. (1996) *The PARCHED-THIRST Model User Guide* (v1.0). Department of Agricultural and Environmental Sciences, University of Newcastle-upon-Tyne, Newcastle-upon-Tyne.

Zadoks, J.C. (1981) EPIPRE: a disease and pest management system for winter wheat developed in the Netherlands. *EPPO Bulletin* 11, 365–369.

ZhiMing, Z., LiJiao, Y., ZhaoQian, W., Zheng, Z., Yan, L. and Wang, Z. (1997) Evaluation of a model recommended for N fertilizer application in irrigated rice. *Chinese Rice Research Newsletter* 5, 7–8.

Zhou, Y. (1991) Potential influences of the greenhouse effect on grain production in China. In: Geng, S. and Cady, C.W. (eds) *Climate Variations and Change: Implications for Agriculture in the Pacific Rim.* University of California, Davis, California, pp. 147–158.

Ziska, L.H., Namuco, O.S., Moya, T.B. and Quilang, J. (1997) Growth and yield responses of field-grown tropical rice to increasing carbon dioxide and temperature. *Agronomy Journal* 89, 45–53.

Appendix: Personal Communications

Name	Position/institution	E-mail
Aggarwal, Pramod K	Centre for Applications of Systems Simulation, Indian Agricultural Research Institute, New Delhi-110012, India	pramodag@giasdl01.vsnl.net.in
Bhalla, Ravi	No. 10, 2nd Cross, Rajajinagar, Pondicherry 605 010, India	bhalla@auroville.org
Bontkes, Tjark Struif	IFDC-Africa, BP 4483 Lomé, Togo	struifbontkes@yahoo.com
Bowen, Walter	CIP/IFDC, Apartado 1558, Lima 12, Peru	w.bowen@cgnet.com
Cadisch, Georg	Dept of Biological Sciences, Wye College, Wye, Ashford, Kent TN25 5AH, UK	g.cadisch@wye.ac.uk
Coventry, Will	Psychology Dept, University of Queensland, Australia	will@psy.uq.edu.au
Dailey, Gordon	IACR-Rothamsted, Harpenden, Hertfordshire AL5 2JQ, UK	gordon.dailey@bbsrc.ac.uk

Name	Position/institution	E-mail
du Toit, Andre	Agricultural Research Council, Summer Grain Centre, South Africa	andre@igg2.agric.za
Fawcett, Roy	Institute of Ecology and Resource Management, University of Edinburgh, UK	roy.fawcett@ed.ac.uk
Hartkemp, Dewi	CIMMYT, Lisboa 27, Apatado Postal 6-641, 0660 Mexico DF, Mexico	d.hartkamp@cgiar.org
Hodges, Harry	Professor of Agronomy, Dept of Plant and Soil Sciences, Mississippi State University, Mississippi, USA	hhodges@ra.msstate.edu
Hoodenboom, Gerrit	Associate Professor, Dept of Biological and Agricultural Engineering, University of Georgia, USA	gerrit@griffin.peachnet.edu
Jintrawet, Attachai	University of Chiang Mai, Thailand	attachai@chiangmai.ac.th
Knight, Jonathan	School of Environment, Earth Sciences & Engineering, Imperial College of Science Technology & Medicine, London	j.d.knight@ic.ac.uk
Knox, Jerry	Senior Research Officer, Institute of Water and Environment, Cranfield University, Silsoe, UK	j.knox@cranfield.ac.uk
Laurenson, Mathew	NARC, Japan	mathewl@narc.affrc.go.jp
Meng Xu	Manager of Crop Model Development (USA/Canada)	mengxu@growthstage.com
Muttiah, Ranjan	Blackland Research Centre, 720 E. Blackland Road, Temple, Texas 76502, USA	muttiah@brc.tamus.edu
Norton, Geoffrey A. (Professor)	Director, Centre for Pest Information Technology and Transfer, University of Queensland, Australia	g.norton@cpitt.uq.edu.au

Name	Position/institution	E-mail
Olzyck, David	US Environmental Protection Agency, National Health and Environmental Effects Research Laboratory, Western Ecology Division, 200 SW 35th St., Corvallis, Oregon, USA	daveo@heart.cor.epa.gov
Pachepsky, Ludmila	Remote Sensing and Modeling Lab, Beltsville, Maryland, USA	lpachepsky@asrr.arsusda.gov
Payne, William	Columbia Basin Agricultural Research Centre, PO Box 370, Pendleton, Oregon 97801, USA	william.payne@orst.edu
Pinnschmidt, Hans	Dept of Crop Protection, Research Centre, Flakkebjerg, DK-4200 Slagelse, Denmark	hans.pinnschmidt@agrsci.dk
Rabbinge, Rudl	Dept of Theoretical Production Ecology, Wageningen Agricultural University, The Netherlands	office@pp.dpw.wau.nl
Singels, Abraham	Agronomy Dept, Experiment Station, S.A. Sugar Association	singlesa@sugar.org.za
Swinton, Scott	Associate Professor of Agric economics, Michigan State University, Visiting economist, CIP, Peru	swintons@pilot.msu.edu
Taylor, John (Professor)	National Soils Resources Institute, Cranfield University, Silsoe, UK	j.taylor@cranfield.ac.uk
Whisler, Frank D.	Professor of Soil Physics, Dept of Plant & Soil Sciences, Box 9555, Mississippi, USA	fwhisler@pss.msstate.edu

Index

All references to individual models are in small capitals, with indications in brackets of the crop or specific feature discussed. Page references in *italics* are to information presented within a table or illustration.